New Series in NASA History

THE MAKING OF A SPACE SHUTTLE CREW

BEFORE LIFT-OFF

Henry S. F. Cooper, Jr.

The Johns Hopkins University Press

Baltimore and London

In Memory of F. Peter Weil

The Johns Hopkins University Press
701 West 40th Street
Baltimore, Maryland 21211
The Johns Hopkins Press Ltd., London

All photographs courtesy of NASA.

Library of Congress Cataloging-in-Publication Data

Cooper, Henry S. F.
  Before liftoff.

  (New series in NASA history)
  Includes index.
  1. Astronauts—Training of—United States.
I. Title. II. Series.
TL850.C66   1987   629.45′009′2   87-2761
ISBN 0-8018-3524-0 (alk. paper)

The paper used in this publication meets the minimum requirements of American National Standard for Information Sciences—Permanence of Paper for Printed Library Materials, ANSI Z39.48-1984.

Contents

List of Illustrations   vii

Foreword   ix

Acknowledgments   xiii

Prologue   1

OCTOBER 4 AND 5, 1984
Launch   7

NOVEMBER 1983 AND EARLIER
Beginnings   15

WEEK OF JANUARY 23–27, 1984
Star Fields and Bathtubs   34

WEEK OF FEBRUARY 27–MARCH 2
A Fire Truck Backs Over the TACAN Shack   47

WEEK OF APRIL 16–20
Surfing into Honolulu Airport   76

WEEK OF JUNE 11–15
Hairy Things   111

WEEK OF JULY 23–27
The Measuring-Unit Dilemma and the
Freeze-dried Computer   138

WEEK OF SEPTEMBER 3–7
Integrated Sims   155

WEEK OF SEPTEMBER 18–21
The Long Sim   184

FORTNIGHT OF SEPTEMBER 20–OCTOBER 4
Countdown   200

WEEK OF OCTOBER 5–13
A Step Up   206

OCTOBER 13 AND AFTERWARD
Smiling and Nodding   241

Epilogue   249
Index   259

Illustrations

Launch of 41-G from Kennedy Space Center	11
Johnson Space Center	11
The Motion-based Simulator	39
The Fixed-base Simulator	39
The 41-G Crew Members at a Press Conference	61
Ted Browder	62
Shannon O'Roark	62
Randy Barckholtz	63
Andy Foster	63
John Cox	64
Mission Operations Control Room	64
The Weightless Environment Training Facility (WET-F)	103
Jon McBride Helping Kathy Sullivan with Her Spacesuit Gloves	103
Kathy Sullivan and Dave Leestma in the WET-F	104
Robert Crippen in the WET-F	104
A Space Shuttle Training Aircraft	131
A Formation of Three T-38 Trainers	131
The 41-G Crew Emblem	134
The Stand-alone Team's Emblem	134
Jon McBride in Zero Gravity	161
Dave Leestma and Kathy Sullivan in the One-G Trainer	161
The 41-G Crew Members before a Training Session	162
Crew Members Depart for Kennedy Space Center	203
The Landing Strip at Kennedy Space Center	203

A Group Photo in Space   211
Kathy Sullivan at a Window in the *Challenger*   211
Kathy Sullivan Floats over Robert Crippen and Jon McBride   212
Marc Garneau and Dave Leestma in the Middeck   212
Marc Garneau and Paul Scully-Power Conducting an Experiment   212
Dave Leestma Reflected in a Water Globule   229
The Shuttle's Arm Deploying the Satellite   237
The Shuttle's Arm Holding Down the Radar Antenna   237
Kathy Sullivan and Dave Leestma in Their Extravehicular
Activity (EVA)   238
Kathy Sullivan and Dave Leestma at the Orbital Refueling
System   239
Kathy Sullivan Inspecting the Radar Antenna   239

Foreword

*T*here is a rush of excitement at launch, an eight-and-one-half minute rocket ride to earth orbit, and an irrepressible eagerness to get to the tasks for which you have trained during the past year. Moments of pre-launch tension are followed by the turbulent and emotion-stirring thrill of launch and then are suddenly transformed into an eternity of wonder—the first glimpse of earth from outer space is spellbinding! Our planet's brilliant and variegated complexion of colors—white, blue, green, brown, red—makes the earth appear glorious from the threshold of space and full of life against the stark blackness of the universe. This opportunity to view our world from earth orbit, this moment of fulfillment, is worth a lifetime of anticipation—and certainly every bit of time spent (sometimes endured) rigorously training for a flight into space aboard our country's space shuttle.

Of course, training for my first flight began, as it does for all astronauts, long before the formal process of flight-specific crew simulations. In fact, training for space flight begins the moment you arrive at the National Aeronautics and Space Administration's Johnson Space Center at Houston. The excitement and initial perceptions of glamour first experienced upon that arrival rapidly give way to a more realistic appreciation of this new profession's requirements and demands. During one's year-long orientation as an astronaut-candidate, one is involved in an enormous day-to-day effort to assimilate as much information as possible on the entire American space program.

Learning about NASA organization and operations, becoming conversant in the endless stream of NASA's technical acronyms, assisting in the design and development of future space systems, participating in the testing of current or proposed space hardware, learning *how* to read abbreviated flight checklists and discovering *why* each step is done in a particular order, involving oneself in hundreds of other scientific issues and engineering tasks—these are all part of an astronaut's introduction to space-flight training. During those first years as an astronaut, one begins to benefit, too, from Astronaut Office camaraderie, bonds of friendship fostered and reinforced by astronauts' mutual interests and concerns—a camaraderie that enables experienced space travelers to pass on to the newcomers the extra bit of knowledge that may ultimately make the difference between a successful rather than an unsuccessful space mission. However, all of this training occurs long before assignment to a shuttle crew.

When it finally occurs, the selection to a specific flight crew is a joyous occasion. It is also the beginning of a new kind of process that will mold the separate crew members into a very special type of team. "Learning the mission" for this team includes forming and solidifying the bond of trust each member shares with the others. The crew members will be spending about a week in space together, but they will be training for that week for over a year. Therefore, immediately after selection announcement, they dig into their flight, determined to become experts in their assigned mission tasks. The breadth of their training is vast.

In addition to the flight simulations, a shuttle crew must become familiar with such dissimilar subjects as photography and rocket engines, astrophysics and electrical repairs, flight procedures and oceanography, public affairs and geology. Career astronauts cannot afford the luxury of specializing in one particular field of knowledge. Instead, the crew members become highly trained generalists. In preparing to be such generalists, they depend heavily upon numerous other people who will take part in this mission—people like the training personnel, the flight controllers, the contractors, the design engineers, and the mission managers. Training for a flight is, by necessity, a highly interactive process that must ensure that both the flight crew and the ground crews are ready to go at launch time. And it is this "go for launch" training experience for the thirteenth shuttle mission which is so vividly recalled by Henry S. F. Cooper, Jr., in his account of the training of our crew—the crew of STS 41-G.

Nearly since its beginning, the National Aeronautics and Space Administration has sponsored a series of narrative histories which document NASA's several programs in aeronautical and space research and development. With the passing of time comes a broadening of perspectives that can enrich our appreciation of an enterprise so singular as the human conquest of flight and the manned exploration of the heavens. In recognition of the need for broadened perspectives on its role in shaping that enterprise, NASA introduces, with this volume, its New Series in NASA History. *Before Lift-off: The Making of a Space Shuttle Crew* documents an aspect of the American space program too often overlooked in more conventional accounts of space exploration. Cooper's account brings to NASA history the sensitivity of a writer whose reporting on the U.S. space program has appeared in the pages of *The New Yorker* for over two decades.

In *Before Lift-off*, Cooper writes about flight from the perspective of an observer privy to specific preflight activities and simulations. His purview is limited to that period of time—one year before launch—in which mission-specific flight training peaks to its greatest intensity. Focusing on the 41-G shuttle mission of which I was a crew member, Cooper surveys and investigates this critical final year of crew training. Reducing to the pages of a book a training process that encompasses thousands of hours, Cooper has made that process appear, in many ways, different from what I experienced. His perspective is nevertheless a valuable one that chronicles the intense training necessary to the forging of a flight-crew *team* ready for almost any conceivable flight emergency and expertly capable of executing shuttle mission tasks. Cooper provides, from an outsider's point of view, a refreshing perspective into the whole crew-training process and condenses into a readable experience much of the drama and the routine, many of the difficulties and the successes, the personalities and the procedures, that contributed to the making of this highly successful shuttle mission.

A writer of great insight and talent, Cooper has captured for his reader a sense of preflight momentum by describing, in terms that both inform and excite, the highly sophisticated and technical training experienced by the crew and its training team. By deciphering the world of fixed-base simulators, EVA training, remote manipulator activity, and integrated sims, Cooper demystifies the dynamic, complex, and often unpredictable regimen of astronaut training. He has also carefully recorded the often disregarded human aspects of flight training, dispassionately revealing crew personalities exposed in the rigidly defined arena of the flight simulator. Wit, patience, determination—even fatigue and flaring tempers—find their way into Cooper's account. This is a historical narrative whose protagonists are in human situations that are very much nonroutine.

Space flight was, for me, an unprecedented experience. In spite of all the training, I believe it is still difficult for any astronaut to be completely ready for a journey into space. Dr. Sally Ride, the first American woman in space and a fellow crew member, once said that in space, "the environment is different, the perspective is different. Part of the fascination with space travel is the element of the unknown—the conviction that it [space travel] is different from earth-bound experiences. And it is!" Getting ready is just the first step before lift-off!

David C. Leestma
NASA Astronaut

Acknowledgments

A number of people have helped me with this book. They include in particular the crew of the 41-G mission, Capt. Robert L. Crippen, Comdr. Jon A. McBride, Dr. Sally K. Ride, Lt. Comdr. David C. Leestma, and Dr. Kathryn D. Sullivan. In addition, there were the two payload specialists who were added to the mission late in training: Comdr. Marc Garneau of the Canadian navy and Dr. Paul Scully-Power, oceanographer. This is the last time I will refer to the crew by their full names and titles, which they never seem to use themselves. The list of credits also includes the team of instructors that stayed with the crew all the way through training and conducted what are known as their "stand-alone" simulations. The team was led by Ted H. Browder and included Shannon L. O'Roark, Randell J. Barckholtz, and William A. (Andy) Foster. For the better part of a year, these nine people formed a tight group, which at intervals tolerated my presence in what was clearly a very intense period of their lives. This book could never have been written without each and every one of them. No crew of astronauts or team of instructors had ever had to put up with a reporter in frequent attendance before, so their generosity is all the more open-handed. The burden of training me, the rankest of rookies, and of explaining what was going on, fell hardest on Ted Browder, the Team Lead. I owe him a special debt of gratitude.

In addition to the astronauts and the stand-alone team, the credits also include the larger team of instructors that conducted the integrated simulations. The "integrated" team was led by Edwin Q. Rainey and included, among many others, Stephen D. Hamm, John E. James, Ronald A. Weitenhagen, B. Keith Todd, Robert Bassham, Carl Keppler III, and James R. Freehling. They also include the flight controllers assigned to the 41-G mission, led by John T. Cox, the flight director. In addition, there are instructors in special subjects, such as

Jim McBride, who was in charge of the crew's extravehicular activity training in the weightless environment training facility, which is a fancy name for a swimming pool; Stephen B. Seus, who taught them photography; and Michele A. Brekke, who taught them how to launch a satellite. They include the flight affairs officer, William R. Holmberg, who was in charge of developing and then applying the flight plan, or crew activities plan, and also their training supervisor, Myron Fullmer, who planned their schedules—even programming where I would fit into them. They include, in fact, every single person mentioned in this book—the entire volume can be regarded as a single, long acknowledgment.

There are many other people who helped with this book who are not necessarily otherwise mentioned between its covers. Eugene F. Kranz, director of Mission Operations, first suggested the idea that crew training might make a good subject for a book. Others include John E. McLeaish and Douglas S. Ward at the Public Affairs Office in the Johnson Space Center, who encouraged this project from the start. There are others in the Public Affairs Office who were wonderfully helpful, including L. John Lawrence, Stephen A. Nesbitt, Billie A. Deason, Janet K. Ross, Andrew R. Patnesky, and Mike Gentry. Having me turn up at monthly intervals for a year, asking to be escorted into the shuttle mission simulators (more difficult to get into than a bunker on the Maginot Line, even for NASA people), did little to enhance their lives.

I must thank Sylvia Fries of the history office at the National Aeronautics and Space Administration in Washington, D.C., who assured me that a thoughtful account of astronaut training in 1984 would be of historical value, even after the explosion of the 51-L mission on January 28, 1986. Finally, I am indebted to my editors at the Johns Hopkins University Press, Henry Y. K. Tom and Carol Ehrlich.

Prologue

Most people are aware of space flights as individual episodes, apparently unconnected, that have definite beginnings, called launches, and definite ends, called landings. They often seem, in the papers or on television, like self-contained expeditions that might as well occur at a moment's notice, like a spur-of-the-minute trip to the country for the weekend. That, of course, is not the case. A space shuttle mission normally takes almost a year to prepare and train for; the preparations and training for all the missions are conducted by different teams whose members are part of the same overall group of astronauts, instructors, and flight controllers at the Johnson Space Center outside Houston, and who all have to use the same few crowded facilities.

A space flight is like an iceberg, most of which is under water—its substructure is an elaborate platform of training and planning that are themselves interconnected like a lattice of crystals; moreover, a series of space flights is like a number of icebergs all emerging from the same continuous underwater block. Obviously, a space flight cannot be understood as an isolated event that takes place in the sky for a few days any more than an iceberg can be comprehended by what ascends above the water, or a building without considering its foundations, or a person without knowing something of his or her history.

As a reporter for *The New Yorker* who has written about the space program for the last twenty years, I have covered a great many aspects of manned and unmanned flights: unmanned missions to the moon, Mars, and Saturn; manned missions to the moon, to earth orbit, and the Skylab space station. I have written about many groups of people involved with space flight: lunar scientists, planetary scientists, flight controllers, crew-systems engineers, and astronauts. The one area I had never written about, largely because it is perhaps the most closely

guarded by the National Aeronautics and Space Administration, was
flight training.

Because astronauts spend the majority of their time in training,
almost all astronaut biographies and autobiographies discuss crew
training to some extent, but always from the vantage of an individual's own experiences; these books or articles almost never deal with
the training of a crew as a subject in itself. There are three well-known
books that go into considerable detail about training for missions in
the 1960s: *The Right Stuff,* by Tom Wolfe, about the experiences of the
seven Mercury astronauts; and *Of a Fire on the Moon,* by Norman
Mailer, and *Carrying the Fire,* by Michael Collins, both about the
Apollo 11 mission to the moon. Crew training, of course, has changed
considerably since the Mercury, Gemini, and Apollo period. There has
been very little of a nontechnical nature written about crew training
for the space shuttle, which, because of the large crews and the reus-
able, winged nature of the spacecraft, is of very different order from
what was done before. Nor has there been any attempt to follow a shuttle crew through training for a mission.

In the winter of 1983, when missions were being launched with
greater frequency and the point was being made that they were becoming "routine," I wrote John E. McLeaish, then the chief of public
information at the Johnson Space Center, to ask if I could be allowed to
follow a space shuttle crew from its formation through its training and
subsequent space flight. I wanted to see how a space flight evolved out
of the training and planning process. Knowing how sensitive NASA
would be on the subject, and how pressed for time astronauts in training are, I said I would be able to get most of the information I would
need from those around the astronauts (their instructors and others
who dealt with them about their mission). I requested a half-hour interview with each astronaut on each of my visits. I said I hoped to go to
the Johnson Space Center for a week every month or so until launch, in
order to get a sense of forward motion in the evolution of the crew and
mission, and also to limit my intrusions on the crew's time. I promised
to be no more of a bother than the proverbial fly on the wall.

I wasn't at all sure my request would be granted. NASA is rightly
protective of crews in training. The space agency itself long ago recognized that a space flight begins with the formation of the crew, if not
even earlier, and everything that happens after that affects the mission in one way or another. A crew distracted from a lesson by a reporter's presence might miss something that could be essential to the

success of the mission later. On top of this, astronauts as a group are shy of reporters, and with some reason. Many of them have had what they consider bad experiences with the press, dating back to the Mercury and Gemini days when they in fact were hounded unmercifully. More importantly, in a profession that demands long hours and acute concentration, their wish to keep the press at arm's length also has survival value.

I heard nothing specific from John McLeaish until the following September, when a number of new crews were announced. Although NASA and the Astronaut Office had by now approved my request, the final permission depended on finding a skipper who would be willing to have me around. One of the new crews, 41-G, McLeaish indicated, looked promising because it was commanded by Capt. Robert L. Crippen, who was a veteran of two previous shuttle missions and was currently in the middle of training for a third; 41-G would be his fourth. Clearly NASA's most seasoned skipper would be the least likely to be thrown off course by having a fly on the simulator wall, and therefore, McLeaish reasoned, he was the most likely candidate for me to consider. Crippen agreed to take me on. It was a privilege never before granted to a writer.

Although Crippen is an easygoing commander, there are two things he is very uptight about, and they are the success of his mission and the safety of his crew. He imposed precise limits on what I would and would not be allowed to do. He did not want me attending in person any lesson involving the crew and an instructor, lest I be a distraction. I could attend classes in the shuttle mission simulators, where the astronauts were not in the same room as their instructors. And I was free to talk to any instructor after any lesson. As I had requested, he agreed to let me talk to each crew member for half an hour on each of my visits—a privilege that was curtailed in the hectic two-month period before launch. I would not be allowed to fly with any member of the crew in a T-38 jet trainer, and I would not be allowed in a simulator cockpit—the crew station—while the crew was being trained there. I hoped that Crippen would become a little freer as time went on, but, like any good commander, he proved a hard man to wheedle. He stuck to the letter of our initial understanding, doing no less and no more, which in fact is a well-tried characteristic of a commander: to deliver exactly what he has agreed, without deviation. Ted Browder, the leader of the team of instructors that trained the 41-G crew in the simulators, once told me that astronauts, in their words, their actions, and

even their thoughts, tend to be extremely economical, wasting little
time on anything that didn't contribute directly to the matter at hand.
They are good at what Browder called "prioritizing"—putting first
things first. Hence Crippen's single-minded dedication to the mission,
allowing no distraction to stand in the way of its safety, is in fact an
essential characteristic of a spacecraft commander.

And Crippen delivered. He placed no restriction whatsoever on
the amount of time I could spend in the mission simulators, the central
feature of crew training, because there I could sit with the instructors—Browder and his team—listening in on the conversations over
radio headsets all I wanted, and the crew wouldn't even know I was
there. Crippen knew that that basically was all I would need, because
the mission simulators are the heart of crew training. No fly on the
wall ever had it so good: I had a comfortable chair. I had my own bug-
ging device—the headset. And I couldn't distract the crew, even if I
wanted to.

In this unprecedented situation, my relations with the crew, although very friendly, were entirely correct and professional—two
marvelous, old-fashioned words necessary to describe a circumstance
essential to the making of this book. Even though I feel I had a good,
even warm, rapport with all of them, I never in any sense became part
of their group, or was admitted to their inner circle. I was invited up to
their office just once, went to only one of their homes on one occasion,
and attended only one party at which they were present. With one exception, I was never encouraged to meet any member of their families.
My instincts were to respect their privacy, which is hard-won and necessary to them. Like Crippen's restrictions, their deep concern with
privacy is a characteristic of astronauts and is part of the story. Despite the restrictions, I was given an experience that brought me as
close to participating in a mission as any writer has ever had or is
likely to have until one actually goes into space—an eventuality that
now seems remote.

This book necessarily is written less from the point of view of the
astronauts than of the team of what is known as the "stand-alone"
simulator instructors who trained the 41-G crew; and in particular of
their leader, Ted Browder, who was the one who took me in hand. Ted,
more than anyone else, was responsible for the day-to-day training of
the astronauts in the simulators, and this book to a certain extent re-
flects his views of this crew's progress, of crew training in general, and
of how a crew is formed. Without Ted's insights, this narrative would
be much the poorer.

Even though this story is about relatively recent events, it is appearing as part of a NASA-sponsored series about the history of aeronautics and space exploration. No one has ever defined when exactly in the past it is that history begins, but any definition should surely contain the idea that in the interval there has been an event, or a series of events, that changes, however subtly, the way in which we look back; an event, or series of events, that may even have changed our present way of doing things. In the history of the space program, the 51-L explosion is clearly such an event, as was the 1967 *Apollo* spacecraft fire or the first manned lunar landing. Certainly the kind of training represented by the 41-G mission exhibits the way NASA instructors did things in midstride of the early shuttle program, before the explosion, and there will be differences when training for shuttle missions starts up again. 41-G was the thirteenth shuttle mission out of the program's first twenty-five, midway to 51-L. To observe the 41-G crew, I was at the Johnson Space Center off and on for a year—from the ninth mission, or the first Spacelab flight, in November 1983, to the fourteenth, 51-A, the satellite-retrieval mission in November 1984—a period during which the Mission Operations Division, which includes the flight controllers and the instructors, and the Astronaut Division were gearing up to support an even more frequent flight rate. There were six flights in 1984 and nine in 1985, and fifteen were planned for 1986. In 1985, NASA was projecting twenty-four flights a year by 1990.

A careful observer will detect in what follows the beginnings of many of the signs of stress that were noted in the Report of the Presidential Commission on the Space Shuttle Challenger Accident, such as the increased flight rate, the tendency to think of shuttle launches as routine, the danger associated with the first minutes of launch, the competition for the limited time in the simulators, the late and erroneous training loads for the computers, the worries about slipping a launch date because of arrears in training, the stress caused by late changes in the payloads, the removal of the ejection seats, and the desire to land at the Kennedy Space Center. I have not pointed out these signs of future trouble in the text; rather, I have reserved such observations for a short epilogue, choosing instead to let the reader look back with me from the new perspective, with both of us sharing the same twenty-twenty hindsight.

In this spirit, I have made remarkably few changes in the manuscript in the light of the *Challenger* accident, though I have made some tense changes, such as changing "Nothing has ever gone seriously wrong with an American launch" to "In October 1984, nothing

had ever gone seriously wrong with an American launch." I made a
few other changes in the interests of not jarring the reader, such as
dropping the very last sentence in the original manuscript, which
read, "His enthusiasm is one quality Ted—and most others at the
Space Center—need never fear losing." Otherwise, I have not changed
any details that were true at the time and that shed light on the way
training was conducted, even if they might seem a little out of place
today. Because I am a reporter and the book was written in the tradition of journalism, there are no footnotes identifying sources. As with
most journalists who write about areas where there are few written
sources, I have relied almost entirely on interviews, and almost all of
these are attributed directly to the speaker in the text.

One feature that will immediately strike a reader of this account
of a shuttle mission is that everyone—astronauts, instructors, flight
controllers—is terribly upbeat about the mission and about space
flight in general. Although the subject of aborts and possible catastrophes came up in training quite often, they were treated with the same
kind of nervous humor that was evident in the ironic use of the word
"kill" as a synonym for "fake out"—there was a feeling that although
what they called "hairy things" could indeed happen, they were extremely unlikely to happen, and even more unlikely to happen to
them. NASA, within limits, has always encouraged a sort of enthusiastic confidence building—it would be unthinkable that people so
close to danger would do otherwise. This mood of excited, enthusiastic
confidence is one of the elements that makes the 41-G mission a subject of history.

Many of the people mentioned in this book, and their colleagues at
the Johnson Space Center, who are looking forward to the return of a
strong space shuttle program, may be dismayed to see the halcyon
years of the program presented as something from the past. Yet history lives if it sheds light on the present. Two things seem clear: One is
that it will be a long time, even after shuttle flights are resumed, before the exuberance described in these pages will return; after the
Challenger explosion, training for shuttle missions can never be exactly the same again. And the other is that, human nature being what
it is, in some not too distant time confidence and exuberance will indeed return to the space program, though perhaps not in precisely the
same form. In both eventualities, the training for 41-G can be looked
back upon for insight and inspiration.

Launch

*I*n front of the Wakula Hotel in Cocoa Beach, Florida, on the evening of Thursday, October 4, 1984, a small electronic sign flashed in ever-changing letters the message that the Wakula, a low cement establishment filling much of a city block, was proud to be part of the space shuttle team. Many people connected with the National Aeronautics and Space Administration stayed at the hotel, which is about five miles from the main gate of the Kennedy Space Center. Then the sign flashed seven names: Robert Crippen, Jon McBride, Sally Ride, Kathy Sullivan, Dave Leestma, Marc Garneau, and Paul Scully-Power. Scully-Power's name took three flashes to get across; the rest took two. These were the names of the crew of the thirteenth shuttle mission, 41-G, which would be launched at dawn the next morning, October 5, 1984. Crippen, the commander, was an old shuttle veteran; this would be his fourth flight. McBride, the pilot, and Leestma and Sullivan, both mission specialists, were rookies; that is, they had never flown in space before. Kathy Sullivan would be the first American woman to go outside the spacecraft on what is called an extravehicular activity, or EVA; she would help Leestma practice procedures for refueling satellites in space. Sally Ride, of course, also a mission specialist, was no rookie, as she had been the first American woman in space when she flew on the seventh shuttle mission in June 1983, a little over a year earlier. As on her first flight, she would operate the manipulator arm to launch a satellite. The last·two, Marc Garneau and Paul Scully-Power, were not, strictly speaking, members of the crew, but passengers, payload specialists added a few months before. Garneau, a Canadian, would do a variety of experiments provided by his nation's scientists, and Scully-Power, an oceanographer, was going along to take advantage of a mission whose main purpose was to conduct various observations of the earth. I knew about the mission—and had come to know the crew and their instructors and other who had helped plan it—because I had visited them at the Johnson Space Center in Houston periodically over the previous eleven months, in order to find out how a mission is planned and how a crew is trained. As space flights were then becoming more routine—plans were in the works to send politicians, history teachers, and even reporters into orbit—I had picked what I thought would be a routine mission to illustrate a process.

At 7:30 p.m., a prelaunch party was in full swing behind the Wakula, where it backed onto the gray Atlantic. None of the astronauts was there. They were already in bed, having had dinner at about four that afternoon; for two weeks, they had been going to bed earlier, and doing everything earlier, so that they would be able to bounce out of bed at 2:30 the next morning to make the 7:03 a.m. launch. The wife of one of the astronauts—Brenda McBride, a slight woman with long brown hair and a tense manner—handed me a beer and told me that she had not seen her husband, Jon, since the day before, when there had been a cookout lunch on a beach near the house at the Kennedy Space Center where the seven astronauts were staying; that afternoon, the families had had a quiet dinner together and said their good-byes. Hence Brenda, and the other spouses and other family members, had not seen the astronauts all day. "Crip put his foot down," she said. "That's his way." She introduced me to her son, Jon, Jr., a stocky thirteen-year-old with neatly combed red hair sweeping over a broad, open, very Irish face, who looked startlingly like his father. He said he was looking forward to feeling the ground shake when the rocket lifted off.

I went in search of Ted Browder, the leader of the team of instructors who had trained the crew for the past ten months in the shuttle mission simulators at the Johnson Space Center, where almost all the preparations for a flight occur, except for the preparation of the actual flight hardware and the launch itself. There were about a hundred people milling around in the court behind the Wakula, making themselves sandwiches out of cold cuts, ministering to two or three barbecue grills, and helping themselves to cans of beer from two big barrels. Browder wasn't hard to find; I looked for the largest knot of people, and then for the person at its center. Browder, who is in his mid-thirties, is a trim man—the result of months of dieting at his console in the simulators—with long, straight hair and gold-frame glasses; his weatherbeaten face seems to have a variety of muscles whose main function is to hold back a grin forever on the verge of bursting forth, for life to Browder is full of hidden amusements, often in the form of pitfalls he and his teammates concoct for astronauts in simulations. Browder enjoys talking, and is good at it; he was the center of attention for anyone who wanted to feel close to the flight because, in the absence of the astronauts, he was the next best thing. "I was just telling the guys about the ascent," he said apologetically, when the last person had melted away and before a new knot of people had begun to collect around him. "I'm going to have to think up some alibi for myself, so that I can get something to eat—like I'm selling life insurance." Clearly, he was having too good a time to do any such thing. He had been the center of attention for a little over thirty hours now, ever since he and his three teammates—Andy Foster, Randy Barckholtz, and Shannon O'Roark, all in their twenties or early thirties—had arrived by car from Houston, a trip that took them twenty hours, nonstop. They had hoped to get there in time for the astronauts' cookout on the beach the day before, but had arrived just too late. "We didn't want to bust in on them when they were with their families," Ted said. Instead, they had gone to their rooms at the Wakula, feeling slightly let down. The astronauts, though, had missed them, and just as Browder and his teammates were locking their doors to go out to dinner, the telephone rang. It was Crippen and his crew on a conference call. "They said they wanted to thank us for all the training," Browder told me. "We wished them luck—I said that was all they would need, because they had all the skill already. Afterward, there wasn't a dry eye among us."

The next day—October 4—Browder and the team took a public bus tour of the Cape. They were keyed up, and talked and horsed around even more than usual. Everything seemed funny. Because of a bad microphone, the tour guide was completely unintelligible—no loss to Browder and the team, of course, since they knew what everything was anyway; but as they passed the various sights, such as the Vehicle Assembly Building, where the rockets and spacecraft are mated, or the headquarters building, where the astronauts would be spending their last night, and heard these places described in a sort of electronic gibberish, they became convulsed with laughter. Then, on the way out of a movie theater at the visitors' center—where they saw *Hail Columbia*, a film about the shuttle in a new, enlarged process called IMAX—Randy Barckholtz, a distinguished-looking ex–air force officer with a thin moustache, stepped on some chewing gum. The team giggled. As he leaned over to wipe it off, his badge dropped on the ground. The team laughed. Then, somehow, he managed to step

on his badge, so that it stuck to his foot. The team almost had hys-
terics; they accused him of wanting to be the Team Leader. ("That's a
sort of running gag we have," Browder told me. "Once, some months
ago, I dropped my car keys; when I bent over to pick them up, I dropped
everything I was carrying; then, when I stood up, I bumped my head.
So, with us, this kind of klutz attack became associated with being a
Team Leader, and whenever anyone else has one, we accuse him of
wanting to be Team Lead.") Things seemed to get funnier as the day
went on. When they went to a fast-food restaurant, Shannon, a slender
blonde in high-heeled sandals, dropped a dispenser full of napkins,
which blew all over the place. She, too, was accused of wanting to be a
Team Lead. In the evening, they calmed down a bit. As he sipped a
beer, Ted told me that nothing truly hilarious had happened for a couple of hours. He and the team had worked themselves into the frame of
mind that everything would be O.K. tomorrow because with Crippen
aboard, the crew would be able to handle anything.

The party broke up early, as everyone would be getting a short
night's sleep. When Browder woke up the next morning, all his worries had returned. He felt the way he had when he had delivered his
son to his first Little League game: a sort of "Now-I've-brought-you-
this-far-and-the-rest-is-up-to-you" feeling, coupled with a sinking
feeling in the pit of his stomach. His spirits returned, though, once he
and his team had joined the astronauts' friends and families; as the
bus for the Crippen entourage, which they were riding in, rolled
through the scrubby land toward the launch site, whenever it passed
one of the tourist attractions, Browder or one of his teammates did loud
imitations of the unintelligible guide of the day before. As the buses
neared the launch site, the team quieted down. They talked in low
voices. "Mostly, we agreed that the crew was ready, that the mission
was in their hands now, and we built up our own confidence by agreeing that we were confident that the crew could handle just about anything that happened," Browder told me later. "We worried that there
was something we hadn't thought of. But then we agreed that we'd
trained them so much, that even if we forgot something, there'd be an
instinct there that would get them through. It's largely this instinct
that we train for—we can't train for every nut and bolt, for everything
that might happen."

In October 1984, nothing had ever gone seriously wrong with an
American launch—although three American astronauts were killed
in a fire during a test on the launch pad in 1967. And on the launch

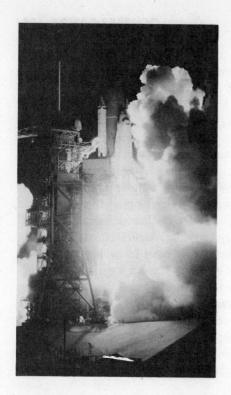

Left, *launch of 41-G from the Kennedy Space Center at Cape Canaveral, Florida, on October 5, 1984, at 7:03 a.m., just before sunrise. The moment culminated almost a year's training and more than a year's planning. Below, a view of the central area of the Johnson Space Center, where missions are planned, trained for, and operated from, showing the duck ponds. Building 4, containing the astronauts' and instructors' offices, is the first building on the right. Building 30, the Mission Operations Control Center, is the second building on the left. The picture was taken from a high floor of Building 1, where the main administrative offices are.*

just prior to 41-G's—41-D's, the previous August—one of the orbiter's three main engines had shut down split seconds before the other two were scheduled to ignite. A fire started afterward on the pad, its flames licking upward around the bottom of the rocket motors. Jon McBride, who had been watching, was puzzled. "That was something I didn't know could happen," he told me later. The fact is that the launch, and the entire ascent phase, is an extremely intricate, tricky, and dangerous business. "Riding a rocket is definitely not routine," John Young, then the head of the Astronaut Office and commander of the first shuttle mission (Crippen had been his pilot then), once told me. "The solid-rocket boosters, bolted on either side of the external fuel tank for the main engines, are the world's biggest solid-fuel rockets; they are state-of-the-art. They have to be 100 percent re-liable—you can't have a nozzle burn through, or have a seal go on you. The same is true for the main engines in the tail of the orbiter. They generate over sixty-four thousand horsepower, but the motors are packed into something like the size of your car trunk. Liquid hydrogen and liquid oxygen are in close proximity to each other. Something could go. And anyone who thinks differently is kidding himself." Young added, "The men and women who fly these things are very brave."

Though the launch was technically not a night launch, it might just as well have been; launch was scheduled for just before sunrise, so almost the entire countdown was in the dark, and although the eastern sky was brightening, there were still stars in the sky. Searchlights surrounded the launch pad, aimed upward at about a forty-five-degree angle, so that the space shuttle and the launch tower were enclosed in a blazing pyramid of light. From the press bleachers, three miles away, I trained a pair of binoculars on the launch complex looking for the orbiter, in which the astronauts were already sitting; about all I could see jutting above the scaffolding was the fat, bomblike tip of the brownish-orange external tank with the tip of one of the white solid-rocket boosters silhouetted against it. There was a much better view on television—about a dozen monitors were at intervals along the foot of the bleachers—because the camera was very near and because it was at an angle that gave better visibility: the orbiter was clearly to be seen, spread-eagled across its raft of solid-rocket boosters and the gigantic external tank. It is a stubby, unsleek craft resembling a boxcar more than an airplane—a boxcar that rides on a pair of huge delta wings. The top and side are covered with white tiles, and the bottom (invisible) with a heavier-duty black tile. A pair of gigantic doors, sixty-five feet long, closes over the back of the fuselage—in orbit, they would open, revealing the cargo bay to space. On either side of the vertical tail are two pods, each containing one of the orbital maneuvering system, or OMS, engines, which give the craft its final boost into orbit. The television view kept changing as different cameras were brought into play; one shot, from underneath, showed the bell-like nozzles of the three main engines clustered below the vertical tail. On either side are the bells of the gigantic solid boosters that would supply the major thrust off the pad.

The countdown continued. At 6:40, the eastern sky began to turn a faint pale orange, silhouetting a flat gray layer of clouds overhead. The pyramid of light encasing the shuttle began to dim. A pond in front of the press bleachers slowly began to turn pink. Mosquitoes began biting reporters; the sound of many slaps could be heard. At 6:58, five minutes before launch, Jon McBride started the three motors for the hydraulic systems that control the orbiter's wing and tail flaps and also aim the rocket motors on gimbals. "We've got three of them up and running," McBride said. Browder and Randy Barckholtz, who were sitting in special bleachers with the astronauts' families, looked at each other and raised their eyebrows; this was precisely the language McBride had used dozens of times in the simulators. They thought he sounded relaxed and in control.

At 7:03, the three main engines in the tail glowed suddenly, followed within seconds by the two boosters, whose brilliance was as dazzling as the sun and (in the comparative dark) hurt the eyes to look at. The light obliterated the pale dawn. Though I had seen dozens of launches on television, none of them remotely approached the genuine article seen live. Television reduces the experience to the size of a small screen, rendering it banal also by repetition and failing to suggest that there might be something more to see, hear, or feel. Not only is the blast of light impossible to reproduce, but it is accompanied by a blaze of sound—the sound and the light are inseparable—and a shaking of the ground as the shock waves make their way across the intervening three miles. While the entire landscape was vibrating, overpowered by the crackling and roaring rocket, Ted Browder looked at Shannon O'Roark and saw tears streaming down her face. "That did it for me," he told me afterwards. That also did it for the rest of the team. It was difficult to realize that there were actually *people* aboard that stupendous controlled explosion—people, of course, whom Browder and the others knew extremely well. (These people, oblivious to the shaking landscape and the burst of light, were calmly ticking off the details of the ascent to Mission Control, just as if they were in one of Browder's simulations.) But the team's feelings may also have had a deeper source. Space flight is otherwise conducted in the absolute silence of a soundless environment, and even the control centers on earth are silent, windowless places. In all this silence, the prodigious blast of a launch is a suitably explosive send-off; it provides the only musical and poetic accompaniment—inarticulate as it may be—of humankind's leaving the earth, a fitting opening chord to going into space, the one tremendous, irrational, earthy event in all the calculated precision and ethereal, otherworldly beauty of space flight.

As the team got itself together, the shuttle climbed on a curly black stream of smoke, the exhaust of the boosters. A thousand, two thousand feet in the air, the lights clustered at its tail were still hurtful to the eyes, the staccato popping of the sound and shock waves still assaulted the ear and shook the ground. The flat layer of clouds overhead concentrated the sound like a low ceiling until the spacecraft disappeared inside it. Two minutes and six seconds into the flight, the two brighter lights at either side of the tail, still visible inside the cloud, sputtered out and dropped away; the giant boosters had done their job. The capsule communicator (or CAPCOM) ticked off various milestones of the ascent, many of them marking abort capabilities in the event of trouble: "Go at throttle up," "Two-engine TAL capability," "Negative return," "Press to ATO," "Single-engine TAL capability," and "Single-engine press to MECO." I was pleased, because I knew what all these meant, and they added up to the successful passage of many hurdles. MECO meant main-engine cut-off, and it occurred about eight and a half minutes into the flight; it was quickly followed by the separation of the giant external fuel tank, which would burn up in the atmosphere over the Indian Ocean. The team remained in the bleachers until MECO. By then the riskiest part was over; the astronauts would make it to orbit safely. The team and the astronauts' families hurried back to the buses to get the jump on traffic. There was no point in sticking around; there's nothing emptier than a just-vacated launch pad.

Beginnings

The Johnson Space Center, where missions are largely planned, orga-
nized, and trained for, is midway between Houston and Galveston. Its
postal address is, of course, Houston, but that is only because the city
managed to snag the Space Center by casting a bit of its corporate pro-
toplasm some thirty miles eastward, so that it is in two widely separated places, a distinction it shares with no other city and with only
one country, the United States. Some cynics might say the Space Center isn't in Houston at all. The neighboring communities include
Nassau Bay and Clear Lake City, as well as a number of others which
started off as housing developments; possibly one of them would be
more appropriate as the Space Center's address. It is in these communities, some dating back no farther than the early 1960s, like the
Space Center itself, that the people who work there live.

Though NASA is perhaps the most visible industry, it is by no
means the only one in what has become a booming part of the country.
Clearly, NASA has helped attract various businesses, among them
aerospace companies; across the four-lane Texas highway that runs
along one side of the Space Center (it is designated NASA 1, to make
the space agency feel at home) are branch offices of the Lockheed Corporation, the McDonnell Douglas Corporation, and the General Electric Company. There are also several large hotels, including a Sher-
aton and a new Hilton overlooking Clear Lake, which connects with
the Gulf of Mexico forty miles away. Along NASA 1 also stretches a
ribbon of gas stations, pizzerias, and shopping malls. Twenty-five
years ago, most of this countryside was flat, scrubby pastureland; cows
wandered over what is now the Space Center, chewing weeds alongside
tumbledown gray sheds or in the shade of rusty iron windmills.

The Space Center looks a little like a college campus, for it is built
around a large quadrangle with duck ponds in the middle, and the

surrounding buildings—including those around a smaller quadrangle farther off—in many cases have columned walkways around
them, as if they were modern versions of Greek temples. Most of them
are made of a white cement that can be glaring in the hot Texas sun.
Otherwise the architecture and landscaping are deceptively relaxing,
in contrast to the sometimes frantic activity going on inside the buildings; in this way too the Space Center is like a college. Nor is this col-
legiate atmosphere accidental; the land the Space Center is on belongs
to Rice University, in Houston, and much of the construction, in the
early 1960s, was done with an eye to the possibility that NASA might
no longer need the Space Center some day, in which case it would re-
vert to Rice, which might want to use it as a campus. Nor is the Space
Center out of place on what might be a college campus. "We run a college here," Eugene Kranz, the director of Mission Operations, which
includes crew training, told me.

To the extent that that is the case, the ninety-odd astronauts are
the students and the 164 members of the Training Division are the
faculty, thus making the Space Center one of the few institutions that
has more faculty members than students. There are many other divisions as well, all of which support manned space flight—the Flight
Control Division, which operates the Mission Operations Control Center during a flight; the Crew Systems Division, which develops hardware for missions; the Medical Sciences Division; the Photographic
Technology Division; the Aircraft Operations Division; the Tracking
and Communications Development Division; the Avionics Systems Division; and many more. All these divisions, however, like the depart-
ments of a college, become involved with training, because they all
have to do with the preparations for missions, which the astronauts
must learn about. Conversely, the astronauts in their training will influence all aspects of mission planning; they have their own ideas and
preferences. "They improve the process as soon as they see it," Francis
Hughes, the chief of the Ascent/Entry Section of the Training Division, told me. "As soon as they show up at the simulators, we are apt to
get a course correction from them—they will have new insights into
what we are doing." Astronaut training is a highly interactive process,
involving every aspect of the Space Center. In this, too, it is like a university, where teaching and research go on side by side, and it is neither easy nor necessary to tell them apart.

To understand how a space flight evolves, one has to go back to its
roots, and these are not in Houston but at NASA headquarters in

Washington, D.C., where the cargoes that are carried into space and that define the missions have always been signed up. In the past, NASA missions started off ingloriously; they spun out a list of cargoes and possible missions called a manifest, in much the same fashion as do the long-haul trips of the vans of a moving company, to which the Space Transportation System, the name of NASA's fleet of four orbiters and their attendant hardware, has sometimes been compared, among others by astronauts, who have referred to it as the Ace Moving Company. In those days, NASA's manifest was almost a living thing, for the number of cargoes was forever changing as flights were flown, as old customers dropped out and new ones appeared, and as the available number of missions changed, depending on the number of orbiters available. A lengthy renovation, such as the *Columbia* underwent in 1984, could unpredictably reduce the number of flights available, and so could a change in the estimate of time needed to service an orbiter between flights. Such problems resulted in periodic massive overhauls of the manifest. The farther ahead one looked, of course, the less firm was any particular flight or its bill of lading. Even after a mission had been decided on and a launch date, a cargo, and a crew assigned to it, a dozen things could change, including its cargo, its crew, its launch date, the landing site, and its duration; indeed, entire missions were canceled just before launch and their cargoes and crews combined with future missions. The reshuffling could upset a year's planning and send ripples far down the manifest.

In the planning of missions, there were two sometimes conflicting principles—first, that cargoes be assigned to flights on something approaching a first-come-first-served basis; and second, that cargoes be grouped together on flights in a way that made some sort of internal sense. As likely an origin as any for the 41-G mission was that NASA's Office of Space Science and Applications (OSSA) in Washington wanted to fly a package of experiments for studying the earth called OSTA 3—OSTA stands for the Office of Space and Terrestrial Applications, OSSA's original name (it was apparently unwilling to change the name of the package at the same time), and this was the third such package. It included the shuttle-imaging radar (a second flight for and a second version of that instrument, abbreviated as SIR-B) and three other instruments, the large-format camera, the measurement of air pollution from satellites experiment, and the feature identification and location experiment. Because the best orbits for studying the earth are the ones with the highest inclination to the equator, the 41-G mission would fly at an inclination of fifty-eight degrees, which would allow it to cover the earth to those parallels north and south of the equator. Because of this useful earth-watching trajectory, other earth-watching items were put aboard, such as the earth-radiation budget satellite, which of course would be launched into an orbit with the same inclination as the shuttle's. Very late, a German experiment called SPARKS, an earth-observing camera, was added to the cargo; it had only recently been signed up for a flight, and therefore should have waited a year or two for its turn, but there was extra space on 41-G, and the mission definition matched SPARKS' requirements. The remaining major payload on the mission, the orbital refueling system, had no particular earth-watching requirement, as its sole purpose was to demonstrate the astronauts' ability to refuel satellites in space. However, NASA was in a hurry to fly this particular demonstration, as it could lead to more business for the shuttle. And it would provide an opportunity for the 41-G astronauts to do an EVA.

*T*he cargo, in turn, had a bearing on the selection of the crew, which was not done until November 1983, about nine months before the original date for the mission's launch, August 1984. The pool of astronauts from which the crew was chosen was in its way as changing and evolving as was the pool of possible missions—not so much with respect to changes in personnel, for the astronaut corps is in that regard remarkably static; but with respect to what the astronauts were doing, and in what state of readiness they were. In November 1983, there were some seventy astronauts, of whom fifteen had been members of the corps since the 1960s, at the time of the Apollo program. Fifty-four had been appointed more recently, in groups, or classes, selected in 1978 and 1980. (A group of seventeen astronauts was appointed in May 1984, and a further group of thirteen a year later, bringing the total at the time 41-G was launched to over a hundred; a few astronauts had resigned.) Of the seventy astronauts present in November 1983, about half were pilots whose job it would be to fly the spacecraft (there would be a commander and a pilot aboard each mission), and the other half were mission specialists, a category of astronauts without piloting experience, new with the shuttle, who would be operating the instruments and launching the satellites it carried. There were eight women mission specialists in the corps. (Two more were added in the class of May 1984, which included seven pilots and ten mission specialists; as

crews get larger, more mission specialists will be needed.) The astronauts range in age from their late twenties to their fifties, with most of
them being in their thirties or forties. About half the astronauts come
from the armed forces, with navy and air force officers vastly outnum-
bering those from the army. Most of the pilot astronauts come from the
military, though some are civilian test pilots. The mission specialists
are a mixed bag. Many of them, particularly the ones that originally
were scientist-astronauts in the Apollo program, have doctorates, frequently in physics, but also in astronomy, geology, and medicine. The
scientist applicants, however, were told before their selection that
their purpose in space would be to conduct other people's experiments
and not their own, a condition that caused a number to drop out; indeed, selection often depended on how willing the interviewers felt the
candidate would be to carry out other scientists' work. Eventually the
requirement for doctorates was dropped, so a few new astronauts have
only bachelors' degrees. On whatever level, though, when the pilots
are considered with the mission specialists, the commonest degree of
all is engineering in one form or another, particularly electrical, aeronautical, or mechanical engineering.

It is almost impossible to tell an astronaut left over from the Apollo
days from a new shuttle astronaut, unless one makes a guess based on
age (almost all astronauts look youthful, but some of the old Apollo
astronauts have gray hair). There are, however, less apparent differences. According to those who know them best, the younger astronauts in many cases have grown up in an era when there has always
been a space program, and consequently they have wanted to be astronauts from a younger age and in many cases have aimed their education at that goal from high school on; they are more apt than their
predecessors to think of the space program as a career instead of a
stepping stone to something else. Consequently, the newer astronauts
may have been better prepared upon their arrival at the Space Center
than the earlier ones. Both groups, though, are quick-witted, decisive,
and ambitious, with most eccentricities selected out; a catalogue of
their extracurricular interests is strong on such things as flying,
handball, jogging, backpacking, auto restoration, and waterskiing. A
very few list reading or classical music. The only one to list numismat-
ics collects Indian-head nickels. "They are great at softball and brush-
ing steaks," one NASA person who doesn't enjoy either pursuit told
me. This is a bit of a put-down. Generally speaking, astronauts are a
bright and lively group to be with; they are deeply involved in what

they are doing, and as is the case with any group committed to a large project, they are enhanced by it.

For the first six months or so after their arrival at the Space Center, newly accepted astronauts—candidates, as they are called during this period, though they are in fact full-fledged members of the corps—undergo a sort of basic training, where they learn the rudiments of the orbiter's systems. Those who are to be pilot astronauts learn to fly NASA's fleet of T-38 fighters, in which they simulate the very steep approach to a runway that the shuttle orbiter travels. Mission specialists get used to the steep descent by riding in the T-38's back seats. They spend about forty hours in the simulators. They train in the Space Center's altitude chamber and in spacesuits, under water, in the big water tank called the weightless environment training facility but known as the WET-F.

When they are through with the candidacy, they move into the pool of astronauts available for missions; while they are waiting, they take on a variety of assignments, all of them related to supporting missions. During the 1970s, between the Apollo-Soyuz mission in 1975 and the first space shuttle mission in 1981, virtually the entire astronaut corps was in this pool, for no flights were being flown—though the first few space shuttle crews, consisting of two men each, were assigned in the late 1970s, and several pairs of astronauts were assigned to five approach-and-landing tests for the shuttle prototype, the *Enterprise,* which was taken up on top of a Boeing 747 and dropped from a height of thirty thousand feet. In November of 1983, however, this situation had changed, as NASA was beginning to build up to what it then hoped would be monthly missions, with crews of at least five astronauts each; about seven crews were in training, representing about half the available astronauts. The other half, those still in the pool, were performing duties—on-the-job-training, really—associated with space flight. They included such tasks as being chase pilots, which means escorting orbiters returning from space down to the runway in T-38s; being shuttle mission simulator liaisons, which means interviewing returning crews about how accurate NASA's simulators had been compared to actual space flight and then improving the simulators; being capsule communicator, which means serving in the Mission Operations Control Center as liaison with the astronauts in space; working on the avionics laboratory, where the shuttle's computers and navigation instruments are tested; or serving as a sort of custodian and editor for the procedures for each mission, which means keeping up to date the manuals and checklists carried aboard the orbiter. The pool, in short, keeps astronauts who are waiting for missions entirely plugged into what's going on; they are, in fact, participants in all the missions, so that when their turn comes, it is a logical next step in what they have been doing all along.

At NASA, crew selection is generally considered something of a mysterious process, as is the case in any selection process where decisions are carried out behind closed doors. Crew selections are made by George W. S. Abbey, the director of Flight Crew Operations, together with John Young, then chief of the Astronaut Office. Young is the astronaut who has been in the business the longest—he flew in *Gemini 3,* in 1965; *Gemini 10,* in 1966; *Apollo 10,* in 1969; and *Apollo 16,* in 1972. He is the only astronaut on active duty who has walked on the moon. More recently, of course, he commanded the first space shuttle mission in 1981, and also the ninth mission—the first to carry the Spacelab— in November 1983. Young is generally considered a straightforward fellow without a shred of mystery to him. Abbey, on the other hand, may also be very straightforward, but he has managed to cultivate an air of considerable mysteriousness. He walks quietly, so he is apt to pop up at an astronaut's elbow when least expected. He keeps out of the limelight and is so unknown to the public—a notable achievement, for someone who is the astronauts' immediate boss—that some of the early shuttle crews felt they should actively publicize him. When he accompanied the STS-5 crew on its way to the launching pad, one of the crew, unknown to Abbey, carried a sign with his name and an arrow pointing to him. Another crew pasted a large photograph of Abbey in the shuttle's middeck so that it appeared on much of their own photography, as if Abbey were aboard. Abbey, a stocky man with crew-cut, almost black hair, has strong ideas about what astronauts should be like, how they should behave, and what they should wear: they should behave conservatively, be somewhat laid back, and dress neatly but informally. Some have complained that astronauts who don't conform to these stereotypes don't get to fly as soon as others.

Mysterious as the process is thought to be, George Abbey assured me that there was nothing difficult or arcane about selecting crews; it was simply a matter of matching the requirements of a mission to the talents available among the astronauts in the pool. Although such a declaration may leave a great deal unsaid, Abbey made it sound as though the logic was virtually inevitable: in the case of 41-G, and of most crews, the first person picked was the commander, who is always a veteran. Abbey and Young chose Captain Robert L. Crippen of the navy, a man in whom they had great confidence, both men having used him as their deputy. Crippen was NASA's leading space shuttle veteran; he had been on two flights already, and he was currently assigned to a third. 41-G, therefore, would be Crippen's fourth flight, but because he was still in training for his third, he would be unable to join his crew until four months before launch, originally scheduled for August 1984. (Both that date, and the date Crippen would join the crew, would slip.) Flying Crippen so often seemingly went against NASA's policy of building up a large pool of experienced astronauts—essential if NASA is ever to achieve a rate of one flight a month—and using Crippen again seemed even more unusual in light of his late arrival. However, as missions were then becoming more frequent, and as the time between them was diminishing, it seemed increasingly likely that a commander coming off of one flight would join his next crew late in its training, and NASA wanted to get some idea of how this situation would work. It was obviously important to have the first commander to try this be a highly experienced one.

Crippen, who at the time was forty-seven, looks a little like Young: both are medium in height (five feet ten inches and five feet nine inches, respectively), dark, and wiry, with sharp, alert faces. Crippen meets Young's and Abbey's standards of what an astronaut should be; he is an experienced flier, has the ability to go to the heart of a matter and make decisions quickly, and is unflappable. Other astronauts like him and look up to him—particularly the younger ones, most of whom seem to want to fly with him; being on a Crippen crew has a definite cachet within the corps. Crippen graduated in 1960 from the University of Texas at Austin, where he took the first computer course ever offered there, and he kept up on the subject later in the navy's Aviation Officer Program. In the navy, he logged five thousand hours flying time in jet aircraft. He joined NASA in 1969 and helped with the Apollo and Skylab programs; while he was in the pool of available astronauts all during the 1970s, in the long years before the first shuttle flight, STS-1, he worked with the technicians who were developing the software for the orbiter's five general-purpose computers, which together formed the data processing system that governed every aspect of the spacecraft. As the computers are the heart of the shuttle—they control all its systems—and as Crippen knew more about them than any other astronaut, he was an obvious choice to sit in the pilot's seat, next to Young, on the first flight. And it was natural that he be chosen soon again to fly as a commander, which he did on STS-7, a flight requiring considerable knowledge of the computer's guidance and navigation software, as the orbiter on that mission would launch a small satellite called the SPAS, fly a number of maneuvers around it (as if it were a buoy) in what were known as close proximity operations, and then retrieve it with the shuttle's remote-manipulator arm. This was also the shuttle's first five-man crew—the first one in which Crippen demonstrated his ability to shepherd younger astronauts into space, for it included four rookies, among them Sally Ride; they were the first of a new generation of astronauts to fly. Crippen was therefore a natural choice to command the STS-13 mission, redesignated 41-C, which had another otherwise all-rookie crew and which would perform some exceptionally complex proximity operations—the rendezvous with, and capture of, the Solar Max satellite, its repair, and its subsequent release. (Under the old flight designations, the initials STS stood for Space Transportation System, and the number is the number of the flight in a straight numerical sequence. In the fall of 1983, though, NASA shifted to a more complicated designation in which the first number is the last number of NASA's fiscal year; the second number is the launch site, with *1* standing for Kennedy and *2* reserved for Vandenberg Air Force Base in California; and the letter identifies the flight within the fiscal year. Most people don't like the new system.) Obviously Crippen has greatly influenced the experience of many members of the astronaut corps, and so, if NASA was looking for a commander experienced enough to risk joining a new crew (41-G) halfway through its training, to see how such an arrangement worked out in practice, they had that man in Crippen. Crippen, Abbey told me, is a "very good resource" that NASA can use with confidence as a sort of trailblazer for new and difficult assignments.

Once Crippen was chosen as commander of 41-G, he took a hand in the selection of the rest of the crew. About five crews were being picked at the same time so that, as with the selection of the cargoes, picking the 41-G crew was part of a larger process. Crippen and Abbey evidently saw eye to eye, because they gave very similar reasons for selecting each crewman or -woman. As part of the process of bringing along future commanders, the pilot on a mission is almost always a rookie, in a sort of leapfrogging of experience. Jon McBride, a navy commander, is an excellent pilot with a good knowledge of aircraft systems; in the navy, he had had considerable experience as an exhibition pilot. "He ought to make a good commander some day," Abbey told me. Sally Ride, of course, had flown with Crippen before, on STS-7, where she had operated the manipulator arm, releasing and retrieving the SPAS. As a satellite was to be released on 41-G, Ride's experience with the arm was one reason for selecting her, but there was another reason as well: while Crippen was assigned to the 41-C mission, Ride, who knew Crippen's preferences, would be a useful stand-in for him.

Kathy Sullivan was picked because she was a geologist, and this, of course, was to be an earth-oriented mission; at the time, she was one of only two geologists in the astronaut corps and the first to fly since Harrison Schmitt, the lunar module pilot on the *Apollo 17* mission to the moon. In the pool, she had been working on EVAs, and there would be one on this mission. Sullivan was also one of the large team of scientists who would be getting information from the shuttle-imaging radar antenna that would be studying a number of sites on the earth. (Abbey stressed to me that the fact that there were two women on the flight was a total coincidence; both of their assignments were logical outgrowths of their abilities and the tasks they would perform.) And Dave Leestma was picked because, as a navy lieutenant commander, he was an expert on aircraft systems, particularly computer software. Abbey thought this would be useful in Crippen's absence, for Leestma could supply some of that expertise. In the pool, he had also been working with the orbital refueling system 41-G would carry. Crippen himself was impressed with Leestma's quickness at electronics; he made a point of telling me that Leestma had graduated first in his class, in 1971, at the United States Naval Academy at Annapolis, where he had been known as "the Wiz." He and McBride had known each other in the navy—they had been in the same squadron, though McBride had been a pilot and Leestma a flight engineer—but, according to Abbey, that was not a reason for assigning them to the same flight now.

Seventy people, the number of astronauts at the time, is smaller than classes at most schools. Many of the astronauts had trained together as candidates, and all of them had spent time together in the pool, where, if they hadn't worked together, they had certainly worked on related projects. More than that, just *being* an astronaut is an intense experience, providing a strong common bond and interest of the sort that makes people accept each other, and get to know each other quickly; it is the sort of galvanizing experience that can cement friendships for life. (For this reason, Abbey told me, it was rare for there to be a falling-out between crew members during a mission: space flight itself provided an intense motivation that overrode any personal irritations that plague other small groups of travelers.) Crippen felt that he already knew his crew well—among other reasons, he had worked with each of them when he had taken over as deputy chief of the Astronaut Office, when John Young had commanded STS-9. Crippen had picked the people he thought would be especially compatible, and he succeeded. Two of the crew told me afterwards that, when the announcements were made, one of their first reactions was pleasure at whom they would be flying with.

Abbey notified the new crew members himself, a task he takes special pleasure in, and which he does in his own arch way. He knows that astronauts in the pool are always on tenterhooks over crew assignments—there is perhaps no more suspenseful moment in their lives, except possibly notification of their selection as astronauts, which Abbey also does himself. When Abbey notified Jon McBride that he had been selected as an astronaut, he phoned him in California, where Jon was at the time, and asked him what the weather was like. "I don't know—I haven't been out," said Jon, who was still in bed. "Well, whatever it's like, it'll be a nice day—I want you to come to work for us," Abbey said. With crew selection, his technique for bringing the good news was just the same—the phone call out of the blue, the innocuous question, the surprise proposition. When he called Dave Leestma, he asked him if he liked what he was doing. Dave said, "Yeah." Abbey asked, "Would you like to do something different—like go fly?" Dave said, "Sure." Kathy Sullivan had just returned from backpacking in the Rockies when her phone rang. Abbey asked her how her work with EVA preparations was going. She said it was going fine. Would she like to go on a real EVA? And on it went.

Of course, whenever the phone call came, and the astronauts heard Abbey's voice on the line, they had a pretty good idea what to expect: they knew that several crews were being selected at that time, and from the kind of pool jobs they were in, they could guess their own chances of being on one or another of them. Astronauts are extremely ambitious, according to several people who have worked with them, and rumors are rife among them. At the same time, astronauts are all fairly laid back, or try to be, and there is no overt competition—which isn't to say that there isn't plenty of it just below the surface. Because Abbey's ways are mysterious, there is no obvious way to get ahead;

how well you know Abbey or John Young is not necessarily going to help. The game, to the extent that it can be played at all, depends more on getting the kind of pool assignments that might lead to being on a crew: being a capsule communicator, it is said, often leads to a flight soon. Similarly, being assigned to working with the manipulator arm may be an advantage, provided there are several upcoming flights that use the arm. But these pool assignments are almost as mysterious, to the astronauts, as the crew assignments are; and the helpfulness of one job over another is by no means certain. Among other things, astronauts keep being rotated from one to another. One thing everyone was sure of in 1983, though, was that with the increased number of flights, everyone's day would come—and it would come again, and again, and again. Meanwhile, though, in the highly charged atmosphere, waiting for Abbey's phone calls heightened the tension, and the mystery.

The crew met together for the first time as a crew at a local watering hole, of which there are a good many not far from the Johnson Space Center. They were a very young-looking group; even Crippen, who at forty-seven was the oldest, is well preserved, with a tanned face and black hair. He wears glasses, but only for reading, an activity he seems to pursue solely in the relative privacy of a spacecraft or a simulator, where he reads checklists. They all wear youthful clothes, sometimes in bright colors—Crippen favors blue T-shirts and brown slacks, while Sally Ride, who was thirty-two, wears boot-cut designer jeans a lot, and Dave Leestma, then thirty-four, whose hair is reddish brown, tends to wear reddish-brown plaid shirts and reddish-brown corduroy boot-cut pants. Kathy Sullivan, thirty-two, goes for a more outdoorsy type of blue jean; anyone can tell at a glance that she is a backpacker and camper. Jon McBride, a big, bluff astronaut of forty, tends to wear greenish T-shirts and khaki pants. Crippen did not keep them very long. He told them what he knew about the mission, and said that while he was away on the 41-C mission, he hoped they would get under their belts the more basic sort of crew training; when he joined them later, they would go on together to the specific training for their particular mission. The training should be fun, he said; if it wasn't, they would know that something was wrong. He ended up by handing out the main areas of responsibilities for each crew member, though he cautioned that he wanted everyone to know enough about everyone else's job so that they could all back each other up. Not surprisingly, these assignments were consistent with why most of them had been assigned to the crew in the first place: Sally Ride to the arm, to launch the earth-radiation budget satellite; Dave Leestma to the orbital refueling system, which meant going on an EVA; and Kathy Sullivan to accompany Dave on the EVA, in addition to being in charge of the earth-watching OSTA package, which included the imaging-radar antenna. Dave would help Sally with the manipulator arm, and he was given another assignment as well: the spacecraft they would be flying, the *Columbia*, which had been used for the first five missions and the ninth (the first to carry the Spacelab), was currently being refurbished at the Rockwell International plant in Palmdale, California, and Dave, whose background was as a flight engineer in the navy, was asked by Crippen to serve as liaison with the factory, to keep the crew abreast of any changes in the ship that would affect their flight. Jon McBride, of course, would be the pilot, and as such he would help Crippen fly the craft. He had a number of other jobs, too: he would be in charge of a big camera called the IMAX, filming aspects of the flight in the new oversized process for a NASA client for a sequel to *Hail Columbia*. Jon would monitor the EVA from indoors, helping Dave and Kathy get into their suits and being in charge of their timeline while they were outside. McBride, Ride, Sullivan, and Leestma would be in charge of inflight maintenance, which meant they would learn to repair equipment that broke down. McBride had two additional jobs: he would be in charge of liaison with the engineers who were making the crew activity plan, the timeline for the entire mission, which necessarily involved every facet of the flight. And, in Crippen's absence, McBride would be in charge of liaison with the Training Division. "I guess I'm sort of second in command," McBride told me once when I pressed him, "but I hate to put it that way, because we're all in this together, and a NASA crew doesn't have a strong chain of command anyway." From beginning to end, it was difficult to see the crew as anything other than a group of five people flying off into space together, so deftly did Crippen and McBride manage matters. "Crip runs things with a very light touch," Ride once told me. "He's relaxed. He doesn't act as a dictator. He rarely gets mad." Once, when I asked him to describe his own job, that of commander, he said, "Shoot, someone has to sit back and watch it all happen."

*T*he astronauts share offices on the third and top floor of Building 4, about halfway down one side of the big quadrangle, across one of the duck ponds from the Mission Operations Control Center. Consequently, astronauts refer to the flight controllers, and the flight controllers to the astronauts, as "those guys across the duck pond." Building 4 is a large rectangular structure that looks less like a college building than like a government one built in a period of great austerity. The Crew Training Division is on the second floor, and the Crew Systems Division, which is responsible for most of the flight equipment, is on the ground floor, so that the people the astronauts need to see most often are near at hand. There can be five or six astronauts in a single small office, and there could easily be twice as many, because an astronaut is hardly ever at his or her desk. There is no particular arrangement of the astronaut offices, though John Young had one of the corner ones overlooking a parking lot, and astronauts on the same crew are put together. As missions are flown and new crews formed, there is a continual reshuffling of the astronauts around the third floor.

The 41-G astronauts (minus Crippen, whose desk was with the 41-C crew) moved into a large office halfway down the back side of the building, two floors above the back door where the food-dispensing machines are. Astronauts often eat on the run, and the back door is the handiest for getting to the simulators. Gradually they made it look lived in. Dave Leestma pasted up near his desk a cartoon of a nearsighted astronomer peering through a telescope at a mobile, just in front of the lens, that had a sun, a moon, and some stars hanging from it. Kathy Sullivan had a document on her bulletin board signed by several members of the Crew Systems Division, welcoming her as the first woman astronaut to enter the vacuum chamber in a spacesuit. And Sally Ride had over her desk a poster promoting the possibility of Pigs in Space, featuring Miss Piggy. Aside from these references, I found almost no special notice whatsoever being taken of the women astronauts, other than a poster that had mysteriously appeared on George Abbey's wall that read, "A Woman's Place Is in the Cockpit," a complaint about the fact that none of the women astronauts were pilots, which some women regard as the last barrier to knock down. Otherwise, the only notice taken of the presence of Sullivan and Ride on the crew that I detected over ten months was a couple of remarks by instructors who felt that, during debriefings after simulations, the women talked more than the men, although it is not clear that they did. "I marvel at how unremarkable their entry here has been," Caroline Huntoon, special assistant to the director of the Space Center, who has had a lot to do with the women astronauts, told me. "In some ways it wasn't easy, because everyone was watching. But most of them had already been in male-dominated institutions, such as graduate schools. The men have been 99.9 percent pure in their treatment of them, and the men have come to really value them."

The first order of business was to meet with their training supervisor, Myron Fullmer, who, if the Space Center were a college, would be their course adviser. In the roughly forty-eight weeks of training ahead, the astronauts would have about a thousand hours of course work, including time in the simulators; they would start off at a fairly relaxed pace, doing perhaps twenty-five hours a week, and end up, for a variety of reasons, doing fifty and even sixty hours a week. There were courses in habitability, which included food management, waste management, and simply knowing where everything was packed in the cockpit or middeck. There were courses in photography—there are twenty-two cameras aboard most space flights, of several different types: television, motion picture, thirty-five millimeter, and seventy-millimeter. There were courses in the care and maintenance of the teletype machine, on which detailed messages are sent up from Mission Control. There were courses in the spacesuit—its care and maintenance, putting it on and taking it off, and working in it. Beyond that, there were courses, taught in the water tank called the WET-F, which the astronauts entered in spacesuits to practice EVAs. None of the courses were taught in what could be called a classroom, with rows of seats facing a blackboard. Spacesuit classes were taught in a room with tables where the various parts of the suit could be laid out and inspected. For the most part, habitability classes, and also photography classes, were taught in what is called the "one-G trainer"— simply a mock-up of the interior of the orbiter, with no attempt to disguise the fact that it was on the earth, in "one gravity." In the same building as the one-G trainer for the spacecraft, there was a one-G trainer called the manipulator-development facility for the arm—a mock-up of the aft control console of the spacecraft overlooking the cargo bay, which could be set up with big balloons representing almost any possible cargo. For the pilots, of course, there were the T-38 aircraft and a Gulfstream that simulated the way an orbiter landed even more precisely; and for all hands a KC-135 that flew in great parabolic curves, providing about thirty seconds of weightlessness each time it went over the top. And then there were the various simulators of the pilot's and copilot's controls and of the various systems they operated, the most important being the shuttle mission simulators.

When the astronauts met with Fullmer, it was his job to study their records (as any course adviser would) to see if they had met all the basic requirements during their candidacy training or while they had been in the pool; sometimes there were gaps that had to be filled, or sometimes Fullmer or the astronaut would feel that a course had been taken so long ago that it would be wise to repeat it—perhaps the astronaut felt rusty, or perhaps equipment had changed in the interval. Even though Jon McBride would not be going out on an EVA, he and Fullmer agreed that he should take some of the basic classes in the WET-F, because, in his role as monitor of the EVA from inside the spacecraft, he should have some sense of what it was like outside. Fullmer excused Sally Ride from taking the basic courses in running the manipulator arm, because she had used it so recently on the STS-7 mission, but she agreed to take more of these courses than she might have in order to help train Dave Leestma, her partner, who would have to start learning from scratch. Crippen went over each astronaut's curriculum with Fullmer, and if he thought that an astronaut had too heavy a load in a particular week, he'd cut some of the classes. If he didn't like a particular mix of things, he'd change it—for instance, one week, when there were four ascent simulations in a row, he asked to have them varied with entry simulations.

The weekly schedules for the 41-G crew looked like those of a group of college students who share most of the same classes but not all, the main difference being that there was almost no time when the astronauts were free. When they weren't in classes or in simulators, they were scheduled for other things, such as flying or taking trips to visit the manufacturers of the satellite or some of the equipment they were taking along. Most days the astronauts would arrive at their offices between seven and eight in the morning, do some desk work, and get certain obligatory meetings out of the way. There was one almost every week with John Young, for the entire astronaut corps. And every Friday morning, at eight, the crew met with Myron Fullmer and one or another of his associates, and with Ted Browder, the leader of their team of instructors in the shuttle mission simulators, for a weekly status meeting, to discuss their progress and any problem areas. The astronauts spent all the rest of their days, until five and sometimes later, in classes and especially in the simulators; the classes lasted one or two hours, but the simulations were usually in four-hour blocks. They would get half an hour off for lunch, if they were lucky. And at night they would lug home briefcases bulging with manuals, handbooks, and checklists.

At the center of their training were the simulators, in which the astronauts would learn to fly the spacecraft and practice every aspect of their mission. Although only the commander and the pilot would fly the spacecraft, the more the mission specialists knew about its systems, the better. The orbiter has a great number of systems—electrical, environmental, propulsive, guidance and navigation, data processing, and so forth, and each of these had to be mastered. At the beginning, the astronauts spent more of their time reading manuals and workbooks for each system; then they progressed, with their workbooks, to a type of computer called a Regency trainer, which was highly interactive—that is, it asked the astronaut questions and responded to his or her answers. It was like a simulator in that an astronaut could call up on a screen a blueprint of, say, the electrical system and turn switches on or off with a touch of his or her finger on the screen.

The Regency trainer, which each astronaut used alone, was a sort of bridge between the earlier exercise books and manuals for the different systems and the single-system trainer, where each astronaut, with an instructor, not only could practice in more detail each system—electrical, environmental, propulsion, data processing—but also begin to work through the checklists. The checklists (some ninety-five pounds of them fly aboard each mission) are basic to flying the spacecraft and pervade the training. For almost everything that happens during a mission, and for almost anything that might go wrong, there is a procedure in the checklists. Most nights, the astronauts pore over them. The next day, they practice them in the Regency trainers, the single-system trainers, and—most comprehensively—in the shuttle mission simulators.

In the mission simulators, all the orbiter's systems are integrated into the same set of computers so that they will play together as they would in flight. The entire crew gets into a simulator and can practice a complete mission, or any part of it, with great fidelity. The crew is by itself, as it would be in orbit; there is no instructor in the next seat (as is the case in the single-system trainer), ready to help out. Instead, the instructors—a team of at least four, under a Team Leader, who are something of an adversary relationship to the crew—sit at consoles in a nearby room. From there, they throw in problems—glitches, nits, anomalies, malfunctions, which they call "malfs," and curve balls of all sorts, usually following a rough script they have lined out previously, or sometimes throwing in a malf or a glitch just for the heck of it. They are, in fact, the closest thing in the space age to gremlins. The mission simulator training begins in the second month and runs through the rest of the training period, almost until the end; the astronauts spend eight or even twelve hours a week in the simulators—more time than they spend doing anything else—and the time increases as the launch date approaches. Most of the time is spent doing what is called "stand-alone" training, in that the crew and the training team are the only elements involved. In the last two months before launch, the stand-alone training gives way to what is called "integrated" training: it is integrated in the sense that the crew is tied in with the flight controllers in the Mission Operations Control Center who will be on duty during the flight.

The mission simulators are the heart of crew training. There is nothing quite like them anywhere else. Simulators, of course, are used in other areas, notably the Link trainers that are used for teaching airline and military pilots; indeed, Link has built the simulator crew stations. Other institutions—in particular the RAND Corporation in California—use computers to set up various kinds of games to teach or do research in warfare, economics, and other subjects. But nowhere are simulators used as extensively or realistically, with as many different groups of people tied in, as at NASA. The reason for this is simple: such a high degree of simulation is not needed for training jet pilots, because the pilot who has learned the basics, can go fly, and from then on can achieve greater proficiency in thousands of hours of flying time. Astronauts get relatively little time in their spacecraft, and almost none of that is concerned with the critical parts, ascent and entry. So if an astronaut is going to learn to fly at all, let alone learn to fly a particular mission, the only way to do it is through hundreds of hours in simulators. As astronauts get more and more experience in space, their simulator time can be cut down, but it can never be cut out. Old astronauts have to maintain their proficiency, and new ones will always be coming along. Even crews composed wholly of veterans will have to be taught to work together. NASA does not ever plan to refly entire crews a second time, as most airlines and air forces do, and this, of course, makes the simulators even more essential. NASA has always preferred to develop flexibility over specialization; instead of permanent crews, which can become stratified and introduce a rigid structure into an otherwise fluid astronaut corps, NASA would rather have a large pool of astronauts who can fly in any combination.

Crippen told me that, if it weren't for the mission simulators, flying in space at all would probably be impossible. Because a spacecraft crew flies only once, and because when it does it has to *be* a crew with all that that implies, the simulators are where the crew is put together. It is where the commander establishes his relationship to the crew, and where the crew establishes their relationship with each other. "A team works well when they all know how each will react in different sets of circumstances," Crippen says. "It's important for me to know that McBride will do certain things under certain conditions. We have to develop mutual trust. And we do this in the simulators. A crew matures a bit when it's been in the simulators. Simulators are a closed environment where we're stressed. You cannot handle malfunctions without working together as a crew, and the simulators allow us to do that. Anyone can pick up a book of malfunctions. But in the simulators, it takes a lot more than book knowledge. The real way to operate, we learn in the simulators."

Star Fields and Bathtubs

*T*he mission simulators are in Building 5, a low, windowless structure next door to Building 4, where the astronauts and their instructors have their offices; unlike Building 4, which is glass and concrete, Building 5 is just concrete, like a bunker. The building is almost as difficult to get into, with guards checking passes, and an automatic security device that decides whether to let you in or not. Astronauts and instructors have been denied entry to their own training sessions. The reason that NASA instituted all this security was the increasing number of military flights that would be flown aboard the shuttle. The mathematical models for some of these missions, including not only their times of lift-off and orbits but also the payloads carried in their cargo bays, were already in the mission simulator computers, and the Department of Defense did not want the wrong people to call up these secret plans on the computer screens.

Inside, a long corridor stretches down the width of the building, with several doors leading off of it. The ones to the left lead to the simulators; Sally Ride and Dave Leestma were just disappearing through one that led to the crew stations. Each is a replica of the orbiter's gray cockpit, in particular of the commander's and pilot's seats and the gray control panels that surround them above, below, behind, and before. Most of the switches for the communications system are on the left, the commander's side, and most of the switches for the electrical system and the propulsion system are on the right side, the pilot's; between them are the computers, and in front of each is the essential instrumentation for the navigation system. The cockpit looks very much like that of a large jet airliner, and is about as crowded with switches, dials, and panels. In the simulators, of course, all the controls, and all the instrumentation, are hooked up to the big Univac and IBM computers that can be programmed to duplicate entire missions or any parts of them. Even the views out the windows—computer-generated graphics that appear on screens—are realistic: in orbit, astronauts see the earth below or fields of stars; out the rear windows, they see the payload bay and its contents. In addition to the sights, they experience the sounds of space flight: the thump of attitude-control jets firing, the hum of a caution-and-warning alarm, the buzz of a master alarm. The crew station is complete with what Ted Browder calls "all the bells and whistles."

There is one crew station for each of the two main simulators, the motion-based and the fixed-base. The motion-based simulator, whose crew station from the outside looks like the disembodied cab of a truck, is used mainly for practicing ascents and descents, and it swivels in all directions—it can tip upright and vibrate, to give the feelings of launch, and it can twist this way and that, to give the feeling of an aircraft banking in the atmosphere on its way home.

Sally and Dave had entered the fixed-base crew station, a larger structure that contains, in addition to the commander's and pilot's seats and controls, the rear control panels overlooking (by means of computer graphics) the cargo bay from which the payloads and the manipulator arm are operated, as well as a rough approximation of the orbiter's middeck down below, where the crew quarters are, including the bunks, the galley, the airlock for going outdoors, and the cabinets full of supplies. The fixed-base simulator does not move; it is used for practicing procedures used in orbit, where there is no sense of motion, aside from acceleration and deceleration. Both the crew stations are in a large white room crammed also with consoles for technicians and dozens of cabinets full of big computers. The mission simulators, with all their computer-generated graphics and models of every system of the orbiter, all interconnected so that they interact together, contain more computers than the Mission Operations Control Center, where real missions are run. The simulators and their computers cost a hundred million dollars, and six hundred people are needed to run them.

Having seen Sally and Dave disappear into the fixed-base crew station, I returned to the main corridor and entered the first of two control rooms—there is one for each simulator. Browder and the rest of the training team were sitting at consoles, listening to and conversing with the crew over headsets—the circuit they were all talking on is called a communications loop. (As the training progressed, other loops, with other groups of people on them, such as flight controllers at Mission Control, could be plugged into this loop, in a complex network.) Each of the control rooms (they are identical) looks like a small section of the Mission Control Operations Control Room. The fixed-base room, which I had entered, is greenish, with a single curving row of five or six consoles at the back. Each has rows of colored flashing buttons and dials and a keyboard, and each has several video screens on which the instructors can call up pages and pages of data supposedly sent by telemetry from the spacecraft but actually generated by the computers, giving information about the pressure or temperature inside various tanks or the voltages in all sorts of electrical or electronic instruments. An extra row of television screens allows the instructors to see the same out-the-window visuals as the astronauts see. There are no windows; the only views are the computer-generated ones. In the mission simulators, as well as in Mission Control, which has no windows either, the result is to narrow everyone's focus to the matter at hand: the space flight, and the training for it. Because of this highly internalized concentration, with no outside influence, it is possible for everyone taking part in a simulation to believe that it is an actual space flight.

The lesson today would be to align the spacecraft's inertial-measurement units, electronic gyroscopes that serve the purpose of a compass in space. The alignment is a sort of three-dimensional compass reading in relation to certain stars called "guide stars" (and not, of course, to any magnetic north), establishing a fixed orientation in terms of roll, pitch, and yaw, against which all movement of the space-craft can be measured. Because the spacecraft would have been launched with the inertial units correctly aligned, the alignments first had to be lost. There are three units, and if one fails, or even two, it or they can be realigned by using the remaining good one or ones as a reference. The instructors, though, wanted McBride to align all three units from scratch, and therefore they had to fail all of them—a spaceship in such a fix, an astronaut once told me, was "dead in the water." Shannon O'Roark, who was in charge of computer training and guidance-and-navigation training, called up on a television screen a catalogue she called a "menu" of possible malfunctions, or malfs; she selected one with her light pen—a sort of electronic pencil she pressed against the screen—and transferred it to another computer-screen page, the malfunction summary page, where she parked it until she was ready to use it. I pulled up an empty chair between her console and Ted Browder's—it was a swivel chair with casters; and because my headset's cord expanded in dozens of spirals, I was able to propel myself up and down the room, to sit by whichever instructor was in the hottest part of the action. This would be my usual seat, and my usual way of operating; it worked very well, except on the most hectic occasions, when one or more of the instructors were also on the move—in which case the expandable cords became as tangled as the ribbons on a Maypole.

The astronauts were just finishing up another problem—a pump for circulating freon through the cooling system had broken—and when they had caught their breaths, Ted nodded at Shannon, who jabbed her light pen at the malf on the screen. Allowing the astronauts to catch their breaths between problems was a luxury permitted only at the start of training.

The inertial-measurement units, of course, are invisible behind the forward control panels, but they operate a dashboard display, a black-and-white ball set into the dashboard in front of the commander and the pilot, which revolves so that it always shows the units' alignment, whatever attitude the spacecraft might be in. One way Jon McBride, who was sitting in the pilot's seat at the right, knew that the alignment had been lost was that an alarm buzzed and a red warning light blinked at him. And if he was still in doubt, the ball in front of him had moved randomly, when the spacecraft hadn't changed its attitude at all. He and the commander had at their right hands a small control stick, which in space fires the thruster jets for attitude control or positional movement, and neither of them had touched his stick. Furthermore, no commands had previously been punched into the spacecraft's five redundant computers to fire any thrusters—the keyboard is between the commander and the pilot, and above it are three video screens for displays from the computer. Almost every command to the spacecraft, whether automatic or manual, passes through the computers.

"Good heavens, we've lost all three alignments!" one of the astronauts in the crew station said, with what seemed like exaggerated alarm. "What will we do now?" Ted frowned. The crew didn't seem to be taking the sim as seriously as he would like, and what was worse, all four astronauts seemed to be talking at once. He told me that one of the hardest things to teach a crew at the beginning of training is how to delegate jobs between them; at the start, there is always a tendency to do things by committee, making decisions by consensus. Sometimes, when crews are stumped, the astronauts just stand around the cockpit and look at each other. At the end of training, Ted said, each individual crew member knows almost instinctively who should be doing what, and when.

In normal flight, the alignments are checked periodically because, like any gyroscopes, the inertial units drift a little. Small errors, under half a degree, can be corrected using the two star trackers, little telescopes on the exterior of the spacecraft sensitive to the light of approximately fifty bright stars scattered around the sky—the guide stars. But when the error is large, as it was now, the astronauts have to do the alignment with a device called the crew optical alignment sight, which they fix in place at a forward or an overhead window, and then use their own knowledge of the constellations to find the guide stars. Jon, who as pilot would sometimes be doing the alignments, told me once that he memorized sky charts while lying in the bathtub at home.

Doing the alignment was a fairly lengthy process, and it was made lengthier by the fact that the astronauts had not yet had all that much experience with alignments in general and finding their way about the heavens in particular. They wanted to get Canopus, one of the guide stars, out the overhead window, and Castor, another guide star, out the front window—a feat, for a neophyte, akin to rubbing one's stomach and patting one's head simultaneously. They managed to get Canopus lined up all right, as it is a very bright star, and then they rolled in search of Castor. Jon did the rolling with the hand controller at the rear of the cockpit, where there was another set of controls for the spacecraft, while Dave and Kathy helped him look out the overhead window for Castor. As Jon rolled the spacecraft (in space, as in the fixed-base simulator, there is no sense of rotation), the computer-generated star fields in the windows changed accordingly. The three astronauts were looking in particular for the Pleiades, for them a milestone on the way to Castor. "There they are!" Kathy said. Dave and Kathy congratulated Jon. Shannon O'Roark shook her head in a vigorous No. Clearly Jon and the others needed more time in their bathtubs with their star maps.

While Jon rolled off once more in search of Castor, Sally Ride, who was getting bored, was trying to call up the computer graphics for the satellite she had launched on the STS-7 mission, the SPAS, in order to play with it with the manipulator arm. The mission simulators are the world's most expensive computer games. Browder was annoyed. "I'm not going to support that," he told some computer technicians over his

The crew station for the motion-based simulator, which looks like the disembodied cab of a truck. It is used for ascent and entry training. At the moment it is tilted upward, for an ascent simulation. The hose is for electrical cables.

The fixed-base crew station, used for orbital training. Since there is no sense of rotational change in space, the station does not need to move. It replicates the entire crew area of the orbiter—not only the flight deck with both its cockpit controls and its rear controls for the payload bay, but also the middeck area where the crew's living quarters are.

headset. As a result, the satellite never appeared. Sally is a lithe, small-framed woman with dark brown hair and an extremely expressive face; she is fast with one-liners and given to a certain amount of mischief. Ted told me once that even among astronauts she is known for her quick intelligence and economy of words and actions—she does both with minimal effort, going directly to the point. She was born in Los Angeles in 1951 and was graduated from Stanford University in 1973; she got her doctorate in physics there in 1978, the year she became an astronaut—she had responded to a NASA help-wanted notice on a Stanford bulletin board. At the Johnson Space Center, she married a fellow astronaut who had joined the corps at the same time, Steven Hawley, a mild-mannered astronomer who was training as a mission specialist for 41-D, scheduled for launch in June 1984. Occasionally he would wander into the simulator control room and ask Ted to tell Sally he wouldn't be home for dinner because he suddenly had to fly to the Cape.

When the 41-G crew had first started in the simulators, Ted had worried that in the absence of Crippen, the only veteran astronaut on the crew, Sally would become what he called "over-factored" into the training, by which he meant the other crew members would tend to look to her for answers rather than develop their own skills. That, however, had not come about, because she herself had evidently understood the danger and compensated for it by letting the others make their own mistakes. She clearly was not going to spill the beans about Castor. When she did help the others, it was always in ways that were at once unobtrusive and to the point. She also served a valuable purpose in telling the others how Crippen liked to do things, and how the simulators differed from the spacecraft—for instance, she had told them that, during landings, they would be able to see more out of the windows of the orbiter than they could in the simulator visuals.

One of the visuals for the star fields flickered and went out, on the monitors over the instructors' consoles as well as in the windows of the crew stations, putting an end, for the time being at least, to the alignment. It was a computer problem that might take some time to fix. "We'll just hip-pocket it now," Ted said to me. "We'll throw other things at them. Everything but the kitchen sink." Time in the simulators was valuable because of the great press of crews to be trained. Consequently, Ted would fill out the remainder of the session with problems he and the team would ad-lib. Shannon, with a stroke of her light pen, broke a switch in the mass-memory unit of one of the onboard computers, so that certain types of information could not be called up on the screens. It was a minor malf; more properly, it was what the team called a "nit" because there was an easy fix: all the astronauts had to do was find a redundant circuit.

Jon went to work on the nit, which would normally be in Crippen's area. Sally Ride, forgetting the satellite, took over as pilot; the astronauts, too, had their own personal redundancy. In the absence of Crippen, they switched around a lot to fill his role and each other's, and in this way the 41-G crew became more versatile than most. When at length Jon announced that he had solved the computer problem, Ted and Shannon were clearly pleased. The astronauts' performances in simulations are not graded in any way; if things go well, the instructors pass on to the next problem without much comment. If things go badly, instructors sometimes punish astronauts by throwing in a worse malfunction; or they might bring up the same problem again on another occasion. In the meantime, the astronauts would be expected to bone up on the problem that caused them difficulty, perhaps even by spending time in the single-system trainer. If a crew member was having serious problems, the team leader might even take it up with his or her commander, but no astronaut has ever been dropped from a mission because of a failure to master a system, for any such situation has always righted itself with extra attention by the astronaut and the instructors.

Without any warning, the visuals of the star fields came back, and the crew returned to the alignment problem. There was, I thought, something rather linear and one-dimensional about the training, at least thus far: there had been an alignment problem. There had been a problem with the computer. There had been a problem with a freon pump for the radiators. And there didn't seem to be any particular connection between them. More complex, interrelated scenarios, Browder assured me, would come later, as the astronauts mastered the basics. More fundamentally, the astronauts didn't seem to be working as a crew but as four individuals—they might almost as well have been alone, each in his or her single-system or Regency trainer. Browder clearly had his work cut out for him if he was to mold these four disparate voices into a unit. He didn't seem dismayed, though. First the astronauts had to be familiar with the individual systems; later they would see how they, and the systems, played together; they were at the transition now. At the moment, in the simulators, they were a step ahead of where they had been in the single-system trainers, because they could work on problems in different systems in the same session. The necessary element was time.

"There's old Pleiades again," Jon said—and this time he was right. Just then, a caution-and-warning light lit on a panel in front of him, and an alarm buzzed. He had been rolling the spacecraft so much, in search of first the Pleiades and then of Castor, that he was running out of fuel for the orbiter's forward thrusters, small jets used for changes in attitudes—pitch, yaw, and roll. There are twelve spotted around each of the two pods for the two maneuvering system engines in the rear and fourteen more around the nose. (These always ran out first, Ted told me, because the rear thrusters could draw fuel—hydrazine— from the far larger fuel tanks for the big maneuvering system rockets on either side of the tail.) Later in the training, he told me, he would simply let the astronauts run out of fuel, to punish them for not knowing their star fields well enough (in the orbiter, the sooner they learned the consequences of their acts, the better), but at this point, he simply wanted them to get the alignment. Accordingly, he arranged for a magic refill of the forward thruster fuel tanks.

At last Jon announced that he had Castor properly lined up and that he had tweaked the alignment. "We've done it!" he said, pleased with himself.

"You guys are great!" Shannon told them, with what might have been the smallest trace of a lack of conviction.

Ted asked if they'd like to try another alignment in the few minutes remaining, but the astronauts said they had had enough.

"And after all the trouble I went to, to get you that extra fuel!" Ted grumbled.

As he was closing down his console, Browder said that in some ways it was an advantage not to have Crippen present, at least during the early phases of training, because it gave the others training they wouldn't otherwise have—for instance, if Crippen had been present, Crippen, not McBride, would have worked on the computer problem. "Jon has never flown in a spacecraft before," Browder said. "He would always feel Crippen's eyes over his shoulders, if he were here. And Crippen is the best. His very presence would make the rest of the crew shy. He's their boss. This way, the crew is much less inhibited; they have the freedom to make mistakes. So it's nice for me now to have these crew members by themselves, and bring them together later with Crip."

He was particularly glad to have a freer hand to work with McBride. Many instructors, when they think of training crews, think in particular of what they call "the pilot pair," meaning the commander and the pilot. They are the ones who have to know the orbiter's systems backward and forward, and not the mission specialists, who technically have no responsibility for flying the orbiter. Operating instruments such as the manipulator arm or carrying out experiments, of course, is complex in its own way, but it does not require the depth of understanding the spacecraft the piloting does. As the pilot pair almost always contains a commander who is a veteran and a pilot who is a rookie, it is the pilot who, in Browder's words, has "the farthest to go," and upon whom the bulk of the training is lavished. McBride was not especially strong in computers, which was not important, Browder said, because computers happened to be Crippen's strong point, and in any case the computers, though located between the commander's and pilot's seats, are generally considered to be on what is called "the commander's side of the house." McBride would have a greater responsibility for the electrical and propulsive systems, which are on his side of the house, and at this point in his training, he was having trouble mastering them. "It's a real bear to keep up with them," McBride told me. "A problem in any of those areas can take a long time to figure out. It's not really difficult—it just takes a lot of work to get familiar with it. You gotta live with it. I take a lot of work home. Every night, I read checklists for the next day's simulations. The only way to learn them is just to hammer through them over and over again."

McBride, I had the feeling, was the sort of fellow who got where he was by dint of hard work more than intuition; but once he had mastered something, he knew it cold. He even looked as if that was the case—he was by far the biggest, most solid-looking member of the crew, with a firm, quiet, and gentle look that was anything but nervous or anxious; in a pinch, one had the feeling he would be solid as a rock. With neatly combed red hair sweeping over a broad, open, very Irish face, he was a grown-up version of his son, Jon, Jr. True to his appearance, he was perhaps the most affable member of the crew, and the most hospitable; several people told me that he was "a real country boy," which one of them defined as "someone who's laid back, openly and genuinely friendly, and comfortable to be around." Both he and his friends make the most out of the fact that he comes from West Virginia; indeed, the governor of that state appointed him as "West Virginia's Ambassador of Good Will to Mankind," a description most of his friends find apt, not only because of his geniality but also because Jon is the sort of fellow who would be likely to know the governor. He grew up in a small town, Beckley, in a house on the side of a mountain. He was graduated from the Woodrow Wilson High School in Beckley in 1960, and was in West Virginia University when the navy recruited him; all the recruiters had to do was put him in an airplane and fly him around a little, he told me, and he knew from that moment that he wanted to fly. He flew sixty-four combat missions in Vietnam. Later, he became an aerobatic pilot for the navy, often at air shows flying a sort of curtain-raiser for the Blue Angels, the navy's crack aerobatic team. He would roar in front of the crowds in a red, white, and blue jet, tipping on its side as he flew by the grandstand, revealing a huge American flag painted on the underside. His superior ability as a pilot was one of the reasons he was picked as an astronaut. However, his job, in the pilot's seat of the orbiter, was as much as anything that of engineer—to master the systems and keep them working, so that Crippen could fly it. "If I had to tell him anything, it would be to concentrate on systems," Browder said.

Earlier that week, McBride had spent an afternoon in the single-system trainer, where he was having a lesson with an instructor who was drilling him in systems he was having trouble with—in particular the propulsive systems the orbiter uses in space: the orbital maneuvering system, the two big rockets in back straddling the vertical tail, which are used for accelerating or decelerating to change orbits or to return to earth; and the reaction-control system, the thrusters used for attitude changes. (The third system, the verniers, small thrusters for minor changes in attitude, were not part of that day's lesson.) The single-system trainers are not in Building 5 but in Building 4, on the second floor, just below the astronauts' offices. Each is a small gray room with plywood walls containing replicas of the pilot's and commander's seats and the control panels and switches that surround them. Because there is only one astronaut, studying only one system, the single-system trainer is necessarily more elementary, and less dynamic, than the mission simulators.

The fact that in the rear of the spacecraft the reaction-control system, which uses the same type of hydrazine fuel as the orbital maneuvering system, can be hooked up to draw fuel from its tanks can cause problems, especially if there is a leak in one of the systems and the astronaut has to find the leak and isolate it: doing this requires a technique most rookie astronauts have difficulty mastering, and McBride was no exception. McBride, who was wearing his usual green T-shirt and tan slacks, sat in the pilot's seat, while his instructor sat in the commander's. First, McBride had trouble finding the correct checklist for setting things up—"The master book sends me to this little booklet, and this booklet sends me back to the master book," McBride said, enunciating a problem that has plagued the readers of manuals and checklists ever since the invention of the wheel. Next, he pressed a button on his hand controller that fired all the control system thrusters at once, to make sure they were all working properly. Because every thruster was matched by an identical one pointed in the opposite direction, there was no motion whatsoever, and the exercise was a little like clearing one's throat. A buzzer sounded—a master alarm: the instructor had set up a series of several problems in the computer that ran the single-system trainer. "Low pressure in the maneuvering system fuel tank," McBride said. There was, in other words, a leak, but whether it was in the fuel tank itself, or in the plumbing leading to it, McBride did not know. There were a number of valves in the plumbing, so that McBride could isolate first one part and then another part of the system until by a process of elimination he found where the leak was; he would know because when he shut the right valve, the pressure in the tank would stop dropping. McBride turned off a valve in the manifold, a chamber where the fuel from the tank was fed into pipes to the two big maneuvering system engines. "That's right," the instructor said. "Now work your way back toward the tank." McBride turned off a valve for part of the thruster system, which was taking fuel from the maneuvering system. The spacecraft began to wobble a bit—the inertial-measurement unit balls in the dashboard spun crazily—with this particular valve off, because several thrusters were out of action; the instructor told McBride to reopen it. He continued to lose fuel. "You'd probably need to think about going to a lower orbit, depending on how much propellant you had left," the instructor said. "You always want to make sure you have enough maneuvering system fuel to get back to earth." This seemed good advice; the lower the orbit, the less fuel needed for getting home. McBride turned off several more valves, isolating a cross-feed line between the thruster system and the maneuvering system, and even the tank itself, before he finally found the leak in another manifold, one that fed fuel to the verniers.

No sooner had McBride solved that problem when there was another master alarm. "It's the right maneuvering system engine—
there's a leak somewhere around its isolation valve," McBride said.
"That's good because the leak is easily isolatable," the instructor said.
"But it's bad, because it means you can't feed that engine from the fuel
tank. I'd suggest you cross-feed fuel from the right tank to the left engine." They talked a bit about cross-feeding fuel from one side of the
craft to the other.

Another master alarm. "It's a forward thruster-system leak!"
McBride said. "Things never slow down around here!" And indeed
they didn't, for during the two-hour session there were some dozen
master alarms. I was beginning to see what McBride had meant by
"hammering through it over and over again"; and, when the lesson
was over, he made an appointment with the instructor to hammer over
again something he still wasn't sure of.

A Fire Truck Backs Over the TACAN Shack

A month later, in the middle of February, there were no particular signs of progress in the simulators. According to Sally Ride, it was wrong to look for them, because that was not the way the training worked. The instructors didn't have the astronauts do blocks of things—ascents, say, or photography, from beginning to end, and then move on to the next thing, the way courses are structured in college—but rather all the different subjects were chopped up and done continuously through training. "Otherwise," she said, "by the time we got to the end of training, we'd have forgotten the subjects we learned at the beginning."

In the simulators, the astronauts were practicing entries from two hundred thousand feet (well within the earth's atmosphere) down to the ground. They were in the motion-based simulator. Dave Leestma was in the commander's seat; in the pilot's seat, Jon McBride was using the hand controller at his right hand to make sweeping S turns for what is called "ranging": when information from the computers indicated that the orbiter had too much speed and might overshoot the landing field, he would make wider sweeps to use up speed, or what astronauts refer to as "energy." On a real mission, this sort of maneuvering is often done automatically by computer, but astronauts generally like to take control as early as possible—and in the simulator, the idea was to give Jon a feel for flying the spacecraft. The stick was the same one used in space for firing the thrusters; on the way down through the atmosphere, the thrusters had automatically been phased out, and now the stick operated the aerodynamic surfaces—the elevons at the backs of the wings, the flap at the back of the fuselage, and the aileron at the back of the tail. The orbiter was both a spacecraft and an aircraft, and some of the controls had to double for both purposes.

By the time I got my headset plugged in, the orbiter, the *Challenger,* was already down to forty-five thousand feet and had survived a number of malfs, including an oxygen leak, a computer problem, and a thruster jet that failed to fire. There had been, evidently, a change in tempo since my last visit, when more time seemed to be spent on one problem; now the problems were more frequent. The *Challenger* was heading for the landing strip at Kennedy Space Center—but as matters now stood, the 41-G mission would not fly aboard the *Challenger,* but aboard the *Columbia;* and since the *Columbia* was not equipped to land at Kennedy, which had only one runway without much margin around it, the 41-G mission was scheduled to land on the broad, dry lake bed at Edwards Air Force Base, where there were several runways. The reason for these discrepancies was that the 41-G mission's computer loads and programs were not ready, nor would they be until a couple of months before launch. Accordingly, the 41-G crew, like all crews, was using the loads for another mission farther along the pipeline, in this case the 41-B mission's, which was scheduled for launch later that month in the *Challenger.* This sort of borrowing worked out all right most of the time. However, the *Columbia,* the oldest orbiter in NASA's fleet, handled somewhat differently from the *Challenger;* for example, it had less redundancy in its systems, and consequently the *Columbia* was less forgiving than the *Challenger.* These and other differences were of great concern to Crippen, who felt that training with *Challenger* loads could be misleading for his crew. Although he was at the most hectic part of his own training for the 41-C mission, he found the time to make his feelings known to John Young, and also to Eugene Kranz, the director of Mission Operations, who was in charge of the Training Division as well as the Flight Control Division; Crippen urged them to dig up some old *Columbia* loads from previous missions. Crippen tried to keep a firm hand on the 41-G mission, even though it was necessarily a firm hand at arm's length. McBride or Ride managed to talk to him a couple of times a week, to keep him abreast of the crew's progress and problems, and all the crew members saw him in the course of their daily rounds—coming in and out of simulators, or in corridors in Building 4.

At forty thousand feet, the instructor sitting just to Ted Browder's right, Bill Russell, who was the instructor in Control (that is, he was in charge of training the astronauts in the propulsion systems and also the hydraulic system for operating the flaps and the landing gear) radioed up to the astronauts that the automatic system for opening the landing gear's isolation valve, which would let the hydraulic fluid into the mechanism for lowering the wheels, had failed. He was making what he called a "ground call"—taking the part of the capsule communicator at Mission Control. Jon opened the valve manually—something, of course, that he did by means of a switch on a console and not by turning on a faucet. Bill Russell would shortly leave the team, to be replaced by Andy Foster; indeed, Randy Barckholtz, the systems instructor in charge of electrical and environmental systems training, had just joined the team since my last visit. Foster and Barckholtz, along with Shannon O'Roark, who was in charge of computer and navigation training, and of course Browder, would stay together until the mission was over. But even Shannon had only joined the team in November. Crippen did not like these changes. "To the crew, we are sort of like a security blanket," Ted told me. "The crew wants to see the same faces. Our team is changing a lot, and I think this bothers the crew—especially as Dave, Kathy, and Jon are all new themselves. They keep joking about changes on our team; I think they need the stability of seeing the same faces."

"You're out of the clouds," Ted said as the spacecraft was down to ten thousand feet, with the runway and the huge Vehicle Assembly Building at the Cape in front of them. Jon put down the wheels. The commander, before the first shuttle mission, had normally done this, until it was found in training that, in leaning forward, there was a risk of upsetting his controls. Even now, different commanders had different preferences for the exact moment of putting down the wheels, and one problem of Crippen's absence now was that McBride would have to find out what his preference was later. When they had landed and the wheels had stopped, Browder informed them that they were twelve thousand feet down the runway—too far.

"Much farther, and we'd be feeding alligators right now," Sally Ride said—the Kennedy strip was all too close to the swampy Banana River. Though the astronauts and the team took the sims very seriously, throwing themselves into them with great conviction as good actors do into their parts, there was also a good deal of joking of the sort actors sometimes go in for—take-outs on their roles.

Sometimes, after each simulation, there was a short debriefing, in which the instructors would explain the problems and criticize the astronauts' performance, and the astronauts could ask questions. Bill Russell explained that he had put the glitch into the automatic system for opening the isolation valve for the hydraulics—something that the astronauts should have noticed but had not, so that Russell, acting as Mission Control, had to prompt them. Jon McBride, who was responsible for lowering the wheels, groaned apologetically, and so did Sally Ride, who as the occupant of the middle seat behind the commander and the pilot, was supposed to act as prompter with the checklists. "It's the sort of thing Crip likes me to check on below Mach 1 [below the speed of sound and hence near the ground]," she said. "I didn't do it this time."

By now, because of all the things that had gone wrong, Sally was awash with checklists, and Dave and Jon had some, too, stuck to the consoles in front of them; in a period of heavy action, the cockpit of either the simulator or the spacecraft can look like a stationery store after a hurricane. They are normally printed on cards or papers that are held together by rings or loose binders, and astronauts often yank out the parts they want. A very few of the checklists have to do with flying actual segments of missions—the entry checklist, for instance, which Jon and Dave had propped up on the computer console between them, fills only one side of a card. Other checklists involve operation of equipment—how to turn it on or off, or how to keep it running. But the bulk of the checklists, the ones they were immersed in now, concern what to do when things go wrong—how to handle malfs, nits, glitches, and anomalies. ("There are a lot of things up there that can eat your lunch—either in small bites or in big gulps," a flight instructor once told me.) For the most part, the lists are written in outline form, with the main topics in boldface type and with the details in smaller, lighter type. The ones for malfs are apt to take the form of branching decisions—if A doesn't work, try B. Sometimes colors are used to follow different pathways for quick reference.

The debriefing over, the instructors at their consoles and the astronauts at their stations began flicking switches, or (as they preferred to say) "reconfiguring" their control panels, so that the orbiter was back at two hundred thousand feet, ready for another entry simulation.

"Here we go!" Ted said as the orbiter began another descent. At one hundred ninety thousand feet, Shannon swung into action with her light pen. "I'm going to break a data bus [a sort of switchboard for information inside one of the computers], for want of anything better to do," she said. As a result, the astronauts lost one of their five general purpose computers. The computers are identical, though they contain some different programs and one or another can be the lead for different phases of the flight, ascent, orbit, or entry. For guidance, navigation, and control, three computers perform all the same calculations as the lead, in order to check on it; if there is a discrepancy, any three can outvote the one that is in error. If there is a problem with any of the four computers, the fifth—the backup—can take its place. Aside from guidance, navigation, and control, which the four on-line computers do together, they divide up the other operations of the spacecraft systems among them. Dave Leestma told me once that he found computers the most interesting system to study (like Crippen, in the navy he had specialized in computerized control of aircraft) because it ties all the other systems togther. "The computers run them all," Dave said. "So if you know the computers, you know the orbiter."

"We've just lost a navigation-state vector," Jon said. A navigation-state vector defines the spacecraft's position in three numbers and the direction and speed in which it is moving in three more numbers. The inertial-measurement units, of course, contribute information to the state vector about attitude and acceleration or deceleration, but it can be refined and updated in a variety of ways, including radar from the ground. The loss of the state vector was a consequence of Shannon's malf, for inertial-measurement unit number 3, which had been supplying much of the information for the vector, was governed by the particular computer that Shannon had failed.

"I never thought of that!" Shannon said to the other controllers, genuinely interested—the orbiter was so complex that a malf could have ramifications even the instructors hadn't considered. In this case, though, the lost state vector was secondary to the failure of the data bus, which McBride had not identified. She wanted McBride to recognize that the computer was down and to engage the backup computer. Shannon was very much the enfant terrible of the team; she loved wielding her light pen and putting in malfs and generally going for the astronauts' jugular. At twenty-three, she was also the youngest member of the team: a native of Tippe City, Ohio, she had graduated from the University of Cincinnati only the year before. Like many of the astronauts, she had been interested in the space program for a long time and had skewed her education toward it: she had majored in aerospace engineering in college, spending a semester at the Johnson Space Center. She had applied to be an astronaut, but had been turned down; she expected to apply again some day. O'Roark had been on the team only a few months. She had big round eyes set wide apart, very white skin, and long blonde hair. She always dressed colorfully; that day she was wearing a yellow blouse; blue slacks that were tight at the top, flaring to bell bottoms; and high heels. Most of her movements were rapid; at the console she was apt to sit quite still, gazing intently at the screens, and when she was putting in a malf, she would purse her lips and lash out with her light pen as if swatting a fly. Then she would sit back and enjoy the results with an all-knowing look that would break into a wide grin if the astronauts solved the problem, and an even wider grin if they made a hash of it.

Sally, in the third seat, told Jon that she thought he should take the inertial-measurement unit off the line.

"No, no!" shouted Shannon, who even more than the other members of the team tended to get caught up in her own scenarios. The astronauts couldn't hear her; to talk to the astronauts, the instructors had to press a button at their belts, part of their headsets. Shannon was upset because the astronauts still had not recognized that the lost inertial-measurement unit was part of a larger computer problem, and she wanted to punish them for their failure to bring up the backup computer. This punishment, of course, would fit the crime, for the most unpleasant consequences would follow such an omission in a real flight. However, Ted restrained her, something that he frequently had to do. "They did the wrong thing, but we're not going to kill them this time," he said to her.

I had seen on several occasions that Ted had to restrain his more exuberant teammates, even stepping in to mitigate a problem as he did now, and I asked him how he knew when to apply the brakes. "We like to give the toughest problems to a crew, to find the threshold at which they can solve their problems, and to push that threshold," he said. "But, we don't want to exceed it. My job is to exercise restraint, to make sure that the team's natural aggressiveness doesn't override training. We don't want to kill the crew, or give them problems they can't handle, or give them too many problems at the same time. It's up to me to provide a sense of balance—to realize when a problem is too much, or there are too many of them, and to put the brakes on. It isn't always easy to know, because management is trying to develop an aggressive type of instructor, and on the team the smell of blood is always close by, and up to a point it has to be." I knew this was true; one of Ted's superiors in the training division, Francis Hughes, had told me that he listened in on the communications loops, and if he found that an instructor was being too tame, or too deferential to the astronauts, he would pull him or her out of the simulators.

Soon the astronauts realized they had been barking up the wrong tree with the inertial-measurement unit, and they switched to the backup computer. Bill Russell, at the Control console, wielded his light pen in order to cause a leak in the pipes leading to one of the jets the astronauts would need during entry. The astronauts didn't notice the leak for a while—there was no alarm, as there normally would be. Then Jon noticed that the pressure was dropping in a fuel tank, and he isolated the leak, very much the way I had seen him do in the single-system trainer. Ted was pleased, because he had discovered the leak despite the fact that his attention had not been called to it by a warning signal. "We put that in to demonstrate to them that, when the backup computer is engaged, you lose a lot of the caution-and-warning system," he said. Over the radio, he told Jon he had done well.

Though I couldn't see Jon, I suspected he was leaning back and stretching, a satisfied look on his face, for it was in that sort of pleased, relaxed voice that he said, "What else can happen now?"

What happened now was that all of a sudden the astronauts lost most of the data on their video screens. Several of their instruments, which when they were working normally caused a white field to appear in a window under their switches, went what the astronauts call "barberpole," which is to say that the white fields were replaced by fields of black-and-white stripes indicating trouble; amber caution lights turned on, and there was the wail of a warning signal. Alarms beeped and buzzers buzzed; all the bells and whistles sounded. The astronauts charged after all the problems at once, solving none of them. Shannon was smiling at her console, her light pen on the table in front of her.

"I never did that before," she said contentedly. "I disengaged the backup computer." If all the instructors collectively were like the gremlins that caused mysterious problems in airplanes during World War II, Shannon resembled them the most. "They should reengage it," she said. Ted sat back expectantly, waiting for this to happen, but when it didn't he put in a ground call telling them to reengage. They did. If they hadn't, they would have missed the runway and crashed, and Browder did not want this to happen. But in the debriefing afterward, he told the astronauts that they had plenty of clues to the fact that the backup system was disengaged, and that they should have noticed it themselves. Shannon leaned back in her chair, her arms folded, a huge smile on her face. She was tremendously pleased with herself.

A natural, even healthy sense of competition existed between the team and the crew. Ted admitted it. "We feel pleased if the crew doesn't get a problem we've spent a long time thinking up, and the crew feels smart if it *does* get it," he told me. "We could kill them every time, but we know that our job, and our reward, is to get them to fly. Training is like a chess game, in that we like to outsmart each other. But when we beat them, or they beat us, we have mixed emotions: when we win, we hope they have learned something, as I think they've done just now. And when they win, we feel humbled, but we also feel pride in teaching them well, the way a chess teacher would feel if beaten by a student. It's a chess game where winning is to have them do well in space." Everyone seemed to thrive on the competition. "We're all on good terms," Dave told me. "It's friendly competition, whenever it crops up. The team, of course, gets more and more devious. We know that they're just trying to prepare us for the day we do integrated sims, and then for the flight. And Ted's good—he's good at keeping the whole show under control."

As he set about reconfiguring his switches for the next run, Ted said that this time the scenario called for a short circuit in one of the orbiter's fuel cells and, farther along, a problem with the ground-based tactical air navigation system, or TACAN, which refines the orbiter's navigation information for landing. He and the rest of the team had worked out the scenarios for that day's runs the previous week—they had a syllabus of material that had to be covered during training, which they were working through, but they also planned the scenarios with an eye to the astronauts' weak points, as well as bringing up again material that had caused trouble before. (The TACAN problem was one such.) The seven training teams all seem to have different ways of operating. On one extreme, there are teams who write out elaborate scenarios and stick to them rigidly; on the other, there are teams that arrive at sims with no scripts whatsoever and improvise everything as they go along. The way Ted sees it, there are dangers in both extremes: to stick too rigidly to a script is to take the life out the situation—Ted needs the flexibility to pursue new situations that might arise inadvertently. The astronauts don't always perform according to the book. Sometimes they make mistakes and should be punished by increasing the heat of the situation; other times they think of solutions that the team hadn't thought of, and the spacecraft is left in some unanticipated configuration of switches, which precludes using the problem the team was going to give the crew next. Or sometimes unexpected ramifications grow out of a situation, and it is nice to be able to take advantage of them. "In all those cases, we have to punt," Ted said. The other extreme, where the teams have no scripts ("They might come in knowing they're going to do ascents, but that's all," Ted once described this method), can work brilliantly, for it leaves the team entirely free to play off the crew's strengths and weaknesses; however, if it fails, it can be a disaster. Members of these teams wear tighter jeans and shinier cowboy boots and walk with a jauntier step than the other instructors, as befits people who live closer to the ragged edge. Ted regards himself and his team as being somewhere in the middle, which is to say they write out scripts but feel free to depart from them, omitting whole sections and ad-libbing new ones. "The team is apt to reflect the Team Lead, and I was never very good at improvising from scratch," Browder told me. "I always like to put some thought into a script—I think I can be more complex that way—and I always like to walk into a sim having a good idea of what we're going to do. But at the same time I always like to feel free to go off on a tangent, to pick up on any new idea that comes along. I might even throw the whole script out, but I always want to *have* one."

Ted was very much opposed to a movement within the training division that the scenarios, instead of being written afresh each time, be canned and used over and over. "We want the liberty to tailor our scripts to suit individual crews," he said. "No two crews are alike." The only person I talked to who seemed to be in favor of reusing old scripts was Eugene Kranz, the director of Mission Operations, who told me that two-thirds of the scripts—in particular the generic material, common to all missions, which was what the 41-G crew was learning now—should be old ones. "We might as well recognize that certain scripts are better than others, and those scripts we should use again," he told me. It is not unusual for senior managers to want to codify things, and for younger ones to want greater flexibility. The disagreement had some of the overtones of the question of academic freedom at universities. As far as I could see, the writing of new scripts, and the execution of them with complete freedom to improvise, generated a spontaneity and enthusiasm among the teams and the crews that would be lost if the scripts were canned. And one of the first things to suffer would be the spirit of people like Ted.

Ted was new as a Team Lead; he had been appointed in November, at the same time as the crew. Before that, he had been at the team's Control console, where Bill Russell now sat, and which shortly would be occupied by Andy Foster. Ted had grown up in Richmond, Virginia, and had gone to Virginia Polytechnic Institute before joining NASA in 1967. He described himself as "the original star-struck kid"; he had wanted to work for NASA since he was a sophomore in high school, when he had visited a cousin who worked for an aerospace firm in Tullahoma, Tennessee. Like Shannon, he had in fact applied to be an astronaut, but had been turned down. He would still love to be one, though he recognizes that being an instructor does not necessarily qualify one to be an astronaut; he thinks he would have the most to offer as a mission specialist sitting in the third seat, reading check-lists, as Sally Ride was doing. Like Shannon, he might some day apply again. "But being where we are lets us live vicariously—we're as close to it as we can be, without actually *being* astronauts," he said. Before joining the team on the Control console in 1981, Browder had held a variety of assignments at the Space Center, sometimes in Mission Planning—he had helped with the abort planning for *Apollo 13*—and he had worked for the engineering and design group that managed the simulators, but he had always hoped to work with the crews. From time to time he had assignments as an instructor. "The simulators are my environment; I function better if I can act dynamically in a real situation," he told me. "I've never been as good as a mission planner. Management must know this, because it keeps putting me back into the simulators."

Ted did not let me attend a script-writing session, lest I inhibit the brainstorming, but he told me what happens in them. Basically he and the other three members of the team gather in what is called a script-ing room—one of two small conference rooms with blackboards on the walls—and each member of the team says what he or she wants to take up in the course of the sim. Sometimes he or she simply wants to check off a square in the syllabus—there is a basic lesson book for each system. Other times it is to exercise the crew in an area where it is weak. "My job is to get the most important things a crew needs to be proficient in accomplished—we don't just script at their weak points," Browder said. "They missed reengaging the backup computer just now, and we'll come back to that in future scripts, but we'll also do other things we never taught them before. As the training goes along, there'll be fewer new procedures; we'll narrow down. As long as my team can show up at the meetings with new ideas, though, we'll script them. At the meeting, we all have to be aware of what each other is doing. If Shannon fails a computer, what will that do to Bill Russell's propulsion system? If Bill puts a leak in, will Shannon's computer failure mean the astronauts can't track it? If that's the case, then either Bill or Shannon will have to withdraw. And the same is true with Randy Barckholtz. If he fails part of the electrical system, what equipment of the others will go down? Sometimes I have an idea for the basic direction the script will go in, or sometimes everyone will just jump in and the script will sort of take shape itself. When the major problems are in place, we fill the spaces between with nits—niggling minor problems that are easily fixed. After the script is done, someone, driving home later, will see a way to improve it, and he'll come to me next morning. I'm very liberal about changes."

Ted showed me the script for the upcoming run. It was a piece of lined white paper with columns for the time, or the altitude above the earth, when the malf would be light-penned into the computer; a short description of the malf itself; and a short description of what it was intended to teach the astronauts. The astronauts had already started down from two hundred thousand feet, the starting point of all that day's runs, and so far nothing had happened.

"If it's so quiet, let's kill an MDM," Shannon said.

"Good idea," said Browder.

It was not in the script. An MDM, or multiplexer-demultiplexer, is a piece of electric hardware whose job it is to relay instructions from a computer to various pieces of equipment in the spacecraft, or even entire groups of equipment, and to receive data from these instruments and hold it until the computer is ready for it. The malf was a nit that the astronauts could fix by recycling a switch—by flicking it back and forth once.

"They're ready for the next one," Browder said; he nodded to Randy Barckholtz, who was ready with his light pen to put in the fuel cell problem. There was less time now between problems for the astronauts to catch their breaths. Randy, a tall, good-looking man in an air force officer's uniform, was new to the team since my last visit; a graduate of the University of Michigan in 1976, he had been a dentist in the air force until 1981, when he had transferred to the space field; he had worked in North Dakota at a radar installation until he had come to the Space Center as an instructor, first in the single-system trainer, and now in the shuttle mission simulators. He was about to leave the air force, but planned to stay on at NASA.

There are three fuel cells in the orbiter which combine oxygen and hydrogen to generate all the spacecraft's electricity (and, incidentally, its water), and Randy had failed fuel cell 2. The three fuel cells fed their power to three main busses—A, B, and C—which were kinds of switchboards into which all the equipment in the spacecraft was plugged; when fuel cell 2 began to die, main bus B began to lose power, and so did the instruments plugged into it. "Oh, look at the amps drop," Sally Ride said.

"We've lost the cabin air vent," Jon McBride said.

"We've lost some instruments on the right side," Sally said.

Ted expected them to do what is called a "bus tie"—connecting the failing bus B to busses A and C, whose fuel cells could supply enough extra power to keep all three busses, and their instruments, going. Instead, they began disconnecting instruments they didn't need from bus B until there was no longer an overload (what is called a "power-down"); moreover, they conducted the power-down rapidly enough to keep pace with the dwindling electricity. Certain needed equipment they shifted to the two good busses. From his console, Browder looked on approvingly. "We didn't think they'd do that," he said. It was one of many examples I would see of the astronauts coming up with unexpected but workable solutions. The sims clearly had a life of their own outside anything that was planned.

While the power-down was going on, Shannon had been at work with her light pen; she failed part of the TACAN. The TACAN's signals come from transmitters housed in what are called TACAN shacks, the final one at the landing strip. The orbiter picks up one station at a time, and the signal is used to give the final update for the navigation-state vector. Dave, in the pilot's seat, suddenly found that, according to the TACAN system, the orbiter was five miles ahead of where the navigation-state vector said it was.

"Super, super, super!" Shannon said, sitting on the edge of her chair; she was always the first to appreciate the unfolding of one of her own scenarios. Clearly she was scenting blood.

Astronauts are predisposed to trust information from the ground more than information generated in the spacecraft. Accordingly, Dave and Jon were inclined to believe the TACANs, which meant that they were approaching the landing strip too fast and should lose velocity by applying the speed brakes, a pair of flaps constituting the rudder that could be made to open into a V at the back of the orbiter's vertical tail. They did this now.

"They've screwed up!" Shannon said, pounding her console with delight. "Come on! Come on! They're going to die!" Instructors and astronauts frequently toss words around like "die" and "kill"; in a sim, of course, these words could be used light-heartedly, as synonyms for "foul-up." Of course, everyone was aware that a similar foul-up in space could result in real fatalities, and it was precisely these that the practice sessions were designed to avoid. Hence the use of these words was ironic, not offensive. A simulated death is always better than a real one, and it is especially benign if it can help forestall a real one.

"What did you do?" Ted asked her.

"It's them! *They* did it! They should have switched to another TACAN station as a check, before they put on the speed brakes," Shannon said.

"I don't understand this," said Jon, who was flying the landing from the commander's seat, and who was beginning to find that matters were not coming out right—he could see the runway and knew he had too little speed.

"You will after the briefing," said Ted, who was getting caught up in Shannon's game.

"I love it! I love it!" Shannon said, as the astronauts got farther and farther off course. "I felt bad when I did this to them before, but I feel super about it now." She had given them a similar problem earlier, and they should have remembered the second time around. "Look! There's the Vehicle Assembly Building, right where it shouldn't be!"

Jon knew now that he couldn't possibly make the landing strip. "Shall we set it down in the water? That'd be the right thing to do," he said.

"Do you want the landing gear down?" Dave, sitting in the pilot's seat, asked him.

"That's not the right way to ditch," Jon said.

Ted ended the sim before they had to ditch; he could freeze the simulation whenever he wanted to, and he did so now. They immediately began debriefing.

"Hi, guys," Shannon said. The astronauts laughed, with a trace of bitterness. "Do you remember two sessions ago, we had two TACAN stations that looked fine and dandy, and I told you to try a third station? And you found the first two were wrong? That's what screwed you up this time. The TACAN station you tuned into was biased—it had a five-mile error. There are cue cards aboard for this kind of problem. The first thing you do is try a different station."

"O.K. Lesson learned there," Jon said.

"And I want you to check with Mission Control when you take TACAN data," she went on.

Ted was beginning to feel a little sorry for the astronauts, and he came to their aid—he felt once again he had to exercise a moderating influence on his team. "In the real world, the ground would call, wouldn't it, if a TACAN station was sending misinformation—if a fire truck had backed over the TACAN shack?" he asked Shannon.

"Yes," she answered, "but maybe the astronauts lost communication with the ground."

Sally Ride, who sensed that Shannon was on the defensive, asked, in a rhetorical manner that turned her question into a statement, "But the ground normally calls about a TACAN error?"

Ted, coming now to Shannon's defense, said, "Yes, but we're not only training you for the real world, we're also training you for far-out things. Say a fire truck ran over the TACAN shack *two seconds* ahead of time, and Mission Control didn't know about it."

The astronauts were dissatisfied by this, and the argument went on all week. They felt they'd been had by the team. "The odds against something like that happening are very great," Jon said to me later. Dave, too, had his nose out of joint. There was nothing new in the argument: early in the Apollo program, astronauts had criticized their instructors for giving them problems they thought were "too far out," and the instructors had obliged by making their problems more "realistic." Then had come the oxygen tank rupture while *Apollo 13* was on its way to the moon—resulting in a risky abort—which at the time one flight controller had told me was "so far out that if we had given it to the astronauts in a simulator, they would have complained that it was not realistic." After that, Kranz, who had been lead flight director for that mission, had told me that training would once more include a few far-out failures. And the far-out problems were more apt to occur now, during the stand-alone training, than later, during the integrated training, when Mission Control was involved and the training was more flight-specific. At present, the training was broader, and therefore Ted had a wider latitude than he would later; furthermore, the stand-alone training (alone, that is, without Mission Control) was meant to teach the astronauts how to fly the spacecraft, and by implication to fly the spacecraft by themselves; hence at this point Ted was not particularly concerned about the lack of a ground call about the TACANs.

"So you see, what Shannon did was not just a dart throw," he said to me.

The members of the 41-G crew at the time of their preflight press conference a month before lift-off. At the top is Dr. Kathryn D. Sullivan, and immediately below her are Dr. Sally K. Ride and Lt. Comdr. David C. Leestma. All three were mission specialists. To the left of Leestma is Comdr. Jon A. McBride, the pilot, and in the front row on the right is Captain Robert L. Crippen, the commander of the mission. The two other men in the front row are the two payload specialists added late in training, Comdr. Marc Garneau of the Canadian navy and Dr. Paul Scully-Power, an oceanographer with the United States Navy. Everyone is dressed in uncharacteristically formal attire, a sure sign that they are in the hands of NASA's Public Affairs Office.

Ted H. Browder, the leader of the team of instructors that conducted the stand-alone simulations for the 41-G mission. He is in the crew station of the fixed-base simulator, at its aft control panels overlooking the cargo bay.

Shannon L. O'Roark, who was the instructor for the data processing (or computer) system on the stand-alone team, sits at her console. Her light pen, with which she deftly inserted all kinds of malfs, nits, glitches, bites, and anomalies, is sitting on her computer keyboard, uncharacteristically idle.

Randell J. Barckholtz, who was the instructor in charge of electrical and environmental system training during the stand-alone simulations, sits in the commander's seat in the crew station of a simulator.

William A. (Andy) Foster, the instructor in Control, which includes the propulsion and hydraulic systems. Although instructors almost never sit in crew stations, as Andy, Randy, and Ted are doing on these two pages, they are vicarious astronauts with (as can be seen here) active fantasy lives.

John T. Cox, the lead flight director for the 41-G mission, standing at his console in the Mission Operations Control Room.

The Mission Operations Control Room during the 41-G mission, with the flight director's console on the left and the capsule communicator's at the right, both in the third row. The console for the payloads is straight ahead, in the second row. The map showing the spacecraft's position is in front, with another screen on the right, on which can be seen the 41-G crew conducting their inflight press conference.

*T*he crew did a great many other things that week. Jon McBride had a lesson in procedures for the spacesuit, which he would have to help Dave Leestma and Kathy Sullivan put on before their EVA, and they all attended a class on habitability—how to get along inside the spacecraft—in the one-G simulator. They learned how to use the waste-management system, NASA's euphemism for a toilet. (There was a duplicate of the space toilet, which the astronauts were encouraged to use, in Building 7, next door to Buildings 4 and 5.) Sally Ride did not feel it was necessary to repeat this class, which she had had at the time of STS-7; her philosophy seemed to be to go to classes where she needed to brush up on what was going on, or where she could help other crew members, and to avoid the classes where she felt confident she knew the material, or where she couldn't contribute to the experience of the others. She was especially loyal about going to sims and the mission simulators, because that was where her experience might be most helpful—though occasionally, if there wasn't that much for her to do (for instance, when the arm, her specialty, wasn't being used), she might slip off to a meeting.

One she spent a lot of time attending was a flight-readiness review of the earth-radiation budget satellite, which she was to launch with the arm. (Flight-readiness reviews are held throughout the training and planning of any mission, the last a week before the launch.) The satellite was a fairly small one that looked like an automobile radiator with wings; its purpose was to compare the energy entering its atmosphere from the sun (known to nonscientists as sunlight) with the amount that was bounced back from the earth into space; among other things, it would tell how much of the sun's energy was retained by the planet and its atmosphere. The first of a group of three such satellites, one of which would go into polar orbit, this one would also study the aerosol content of the upper atmosphere. It was managed by the Goddard Space Flight Center at Greenbelt, Maryland, and there were a great many Goddard people at the meeting, as well as managers from the Johnson Space Center. All the astronauts were there, with the exception of Robert Crippen, of course, and so were some of their instructors.

The meeting illustrated well how intricate planning a NASA mission is. Sally was to release the satellite very early in the mission, on the fifth or sixth orbit. Before that happened, the Goddard flight controllers, who would remain at Goddard during the mission in what

NASA calls a Remote Payload Operations Control Center (it is "remote" in that it is not at Houston), would have to check out the satellite; they would have only three or four orbits, or about six hours, to do this. Originally, they had planned their checkout checklist on the assumption that there would be available two communications satellites—called tracking and data-relay satellites—on opposite sides of the earth so that they would provide virtually uninterrupted communications between the orbiter and the ground; but the previous summer they had been told there would be only one. For half of each orbit—over the Western Hemisphere—there would be constant communications; for the other half, communications would be intermittent, conducted through a number of ground stations. This meant that the Goddard engineers had had to reduce their checklist to fit into just half of each orbit, when there would be coverage by the relay satellite. At the meeting, everyone had been impressed by the way the Goddard people had condensed their requirements. Still, there were a number of loose ends to be settled and conflicting interests to be reconciled.

One was settled now: the earth-radiation budget satellite people would have liked the spacecraft to go into orbit three hundred miles above the earth, which would make it easier for them to deliver the satellite to its final orbit three hundred sixty miles up; they could conserve more of its fuel for attitude changes and maneuvers later on, thus giving it a longer life. Crippen therefore had been looking into doing what is called a direct insertion—burning the shuttle's main engines longer so that the spacecraft would go into orbit without using the smaller orbital-maneuvering system engines, which later would boost it higher. A direct insertion had never been done before, and he had been intrigued with trying one. But then the scientists and engineers for the OSTA package, which contained several earth-watching instruments, including the imaging radar, had pointed out that if the spacecraft went up to three hundred miles, it would be difficult to get back down again to the optimum altitude for the instrument package, which for the best resolution had to be below one hundred forty-eight miles, and preferably at one hundred twenty-one miles' altitude. More seriously, if the spacecraft went up to three hundred miles and then dropped down to the lower orbits, it would never get what is called its "phasing" right—that is, it would not be able to pass over the right targets at the right times, or else it would pass over them obliquely instead of directly; the spacecraft could only get it phasing right if it never went higher than two hundred miles. Jon McBride now reported to the meeting a conversation he had had earlier with Crippen, who had reluctantly decided to give up the direct insertion and stay lower down—because he didn't want to ruin the experiments in the OSTA package, particularly if the satellite was capable of getting itself up to the higher orbit with a little expenditure of its own fuel. This was a fundamental decision that shaped the entire timeline; it had to be nailed down early, because many other decisions depended on it.

One such was a matter that came up next: how long would the spacecraft stay at its higher two-hundred-mile altitude before dropping down to the lower orbits where the OSTA package could function best? The sooner the better, the instrument-package people present felt. This indeed could well happen if the satellite got off on schedule on the fifth or sixth orbit (there would be two opportunities in case it couldn't make the first one), but what if it didn't get off then? The satellite people, of course, wanted the option of being able to stay at two hundred miles as long as possible, to make sure they got the satellite off, but the OSTA people wouldn't hear of doing that. How low could the shuttle be and still manage to launch the satellite to its three-hundred-sixty-mile destination? The satellite people would go down as far as approximately one hundred fifty miles, but no farther. As it happened, one hundred forty-eight miles was the upper limit of where the orbiter could get its phasing right for the OSTA package. After some further horse trading, as NASA people like to call such negotiations, it was decided that the orbiter, if it failed to launch the satellite on the first day, would remain for one more day at the higher orbit for a backup launch attempt on Day 2, and then it would drop down to one hundred forty-eight miles, where the OSTA package could begin functioning, and where there would be a third attempt to deploy the satellite. After that, the spacecraft would drop down to OSTA's preferred altitude of one hundred twenty-one miles.

There was some discussion of the actual deployment itself, which would be done by the manipulator arm—a fifty-foot boom that resembled a human arm in that it had a shoulder joint, an elbow, and a wrist in more or less the right places, but instead of a hand there was a cylinder, open at the end, that served as a grapple. When Sally Ride, who would be doing the deployment, discovered that the satellite was going to be stowed in the cargo bay in such a way that its grappling pin was on the side away from her, she complained that she wanted to be able to see the pin; otherwise she would have to depend entirely on a television camera mounted on the arm's wrist joint. Obliging Sally

would mean turning the satellite around in the bay. That meant turning the orbiter around for deployment, so that when she released the
satellite, it would be in the right orientation for firing its own rockets
to higher orbit. In view of all this, did Sally still want to be able to see
the pin? She did.

They then discussed the motions of the different arm joints to
bring the satellite up to the deployment attitude; Sally, who had been
practicing with the arm in various simulators, was able to tell them
what positions should be avoided on the way up so that one or another
of the joints didn't lock, in what is called a singularity—something
that happens when the arm makes the wrong series of movements.
She was also able to report that Crippen was insistent that when the
satellite was released, the arm position must be configured so that the
deployment would be visible through the overhead windows. When he
fired thrusters at the rear to slow the orbiter to drop away from the
satellite, he wanted to be able to see it, to make sure that he didn't hit
it with the orbiter's tail.

Planning and training go hand in hand, with the planning defining
what the training would prepare for, and with the training turning up
problems in the flight plan or finding better ways of doing things. The
person who kept track of all the new ideas and requirements and made
sure that everything fitted together was William R. Holmberg, the
flight activities officer who was in charge of building the crew activity
plan, which was the schedule for exactly what would happen from
launch to landing. The plan is to a space flight what a budget is to a
government, and its formation involves about as much thought and
struggle, because it defines what, when, and how things will be done.
It *is* a budget, with time being the commodity rationed; and in a space
mission, time is always in short supply. The plan was constantly being
assembled and pulled apart as requirements changed; like NASA's
flight-cargo manifest, it was almost a living thing that was in a constant state of evolution. At the moment, the mission was scheduled to
last ten days and would include (in addition to the satellite deployment and the taking of information by the OSTA instrument package)
an EVA on the fifth day, when Leestma and Sullivan would set up the
orbital refueling system; the times of the hundred or so attitude
changes for aiming the imaging radar antenna at various targets;
times for various orbital maneuvering system burns to improve phas-

ing for the instrument package; times for eating, sleeping, and exercising; times for carrying out medical experiments; for switching on and off other experiments in the cargo bay, such as the German SPARKS and several small, privately owned packages called getaway specials; time for photography by crew members; and dozens of other things. Payload requirements are the main drivers of the timeline, but there are others as well. From past experience, Bill Holmberg knew that Crippen had certain idiosyncrasies and preferences; for instance, he was a strong believer that the entire crew should take meals together, which meant that Holmberg should steer away from planning any kind of activity, even involving only one astronaut, at mealtime. Crippen also believed that the hour before bedtime should be time off, when there would be as little planned activity as possible, and that nothing whatsoever should be scheduled during sleep periods. (Other commanders were less strict about meals, encouraging their crews to eat whenever they felt like it, and they were more flexible about doing work before and after bedtime.) Building the crew activity plan involved considerable horse-trading between Holmberg and his assistants, the crew, and all other interested parties. Jon McBride, who was the crew liaison with the planners, met Holmberg at least once a week— —conveniently, his office was a few feet from the crew's on the third floor of Building 4, suggesting the close connection NASA felt was necessary between those who plan missions and those who fly them. Once, Holmberg scheduled seventeen hours of work in one day, cutting into the presleep period; Jon said that would be O.K., provided only fourteen hours of work was scheduled on the next day. He relayed a complaint from Kathy Sullivan and Dave Leestma that Holmberg had scheduled too little time between their return from the EVA and their next tasks—they thought they should have a little rest period in between. Dave had also noticed that when he was at the most critical part of the work with the orbital refueling system, after he had connected the two tanks for the refueling experiment, the spacecraft would not have coverage by the relay satellite—there was a requirement that the ground be able to read the pressures in the tanks, lest there be a leak and the two astronauts get doused with hydrazine, a highly toxic and explosive fuel. Crippen, who had many reservations about experimenting with hydrazine, backed him up. Holmberg had to make an adjustment in the plan. Dave, who in many ways was as sharp as any of the crew when it came to identifying problems and suggesting solutions, also discovered that the checkout of the arm prior to deploying the satellite came right at mealtime. Holmberg put the checkout a little later and made the meal a little earlier. One running battle that the crew—supported by Crippen—was having with Holmberg was to keep the plan simple, by not writing in every detail. "Crippen wants to make it simpler for us," McBride told me once. "He wants to have it freer—he doesn't want it to specify when or who will do every routine chore." The astronauts were pushing Holmberg simply to make a list of the routine chores and let the crew do them when they could.

Sally Ride and Dave Leestma spent two four-hour sessions practicing with the manipulator arm in preparation for releasing the earth-radiation budget satellite. The first of the sessions, where I caught up with them, was in the manipulator-development facility, a life-sized mockup of the cargo bay, the arm, and the control console at the rear of the cockpit from which the arm was operated. The manipulator facility was in a giant hangar of a room in a building beyond the central quadrangle and duck ponds of Johnson Space Center. In it, in addition to the facility, was what was called the one-G simulator of the interior of the cockpit and the middeck below, where the astronauts had their habitability and photography training and learned to do such things as change the canisters of lithium hydroxide that purified the cabin atmosphere. The same huge room also contained a simulator called the air-bearing table, where large objects could be pushed around in two dimensions, as though they were weightless. The room had a bad echo that encouraged hushed voices, as is the case with big factories, which the room, with moveable cranes near the ceiling, resembled.

A wooden staircase led to a landing at the rear of the facility. Sally and Dave, both wearing blue jeans, were standing at the rear console mockup, along with a couple of instructors, Jim Arbet and Keith Todd; Dave was unlocking the arm from the side of the cargo bay, its resting place. It did not look much like the one used in space: that arm, which weighed nothing in orbit, was very slender for all its fifty-foot length; its joints were hardly noticeable, unless they were bent. It was the arm of a weakling. On the ground, however, the arm had to be thick and muscular just to lift itself; the shoulder joint, which in space was only four inches in diameter, here was thirty-two inches across, to provide the extra heft needed to hoist what amounted to a fifty-foot metal beam and hold it at any angle. It looked like a telephone pole—one that could bend in several places. To cut down as much as possible the need for muscle, the payloads for the arm to pick up where balloons molded in the correct sizes and shapes, but otherwise weighing almost nothing: they were filled with a light gas that made them virtually buoyant in air. (In space, the spindly, weakling's arm could toss about objects that on earth weighed many tons, despite the fact that it could exert what amounted to only forty pounds of push, pull, or lift.) Occasionally some of the payload balloons, which were tethered to the arm, got away; and when that happened they were retrieved by workmen on high ladders—unless the balloons happened to hit a light on the ceiling, in which case they burst and dropped to the floor. As Dave, using a hand controller on the console, undocked the arm and raised it, its machinery whirred; the powerful arm worked hydraulically, unlike the arm in space, which had electrical motors to move each joint. Again unlike the arm in space, this one jerked erratically; when Dave stopped it by applying brakes (there was one for each joint), it shimmied back and forth. Though it didn't seem to behave much like the real arm, Sally Ride told me later that it was useful for practice in tandem with the computer-generated one in the mission simulator, where she and Dave could hoist computer-generated payloads in a computer-generated cargo bay, as if they were playing a video game. The mission simulator with its graphics did not give them a sense of the real hardware; for instance, if they hit a payload (or, for that matter, the side of the orbiter) with the arm, the graphics allowed it to pass right through the obstruction. Here, in the manipulator facility, if they banged into something, either the arm would stop, or they would put a dent in whatever they hit—in the case of a balloon, of course, it would pop. So despite its limitations, the facility gave the astronauts a different sort of look at what the actual flight article would be like.

Dave next began bending one joint at a time, to make sure they all worked properly, a little the way a person waking up in the morning might stretch an arm and bend it, first at the shoulder, then at the elbow, before getting up to prepare for the day. He did this in what is called the "single-joint mode"; the arm could be operated that way, which was useful in case one of the joints had a malfunction, or all the joints could be operated together to perform a task by using the hand controller, which resembled the hand controller for operating the spacecraft itself. Indeed, Sally Ride once told me that operating the arm (or "flying" it) was a little like flying a spacecraft. The arm could also be operated in an automatic mode, by which at the press of a but-

ton the grapple could be made to move to any one of two hundred pre-
set points programmed into the computer; from any one of those
points, the astronauts could take over manually to do a particular job.

When he had finished checking out the arm joint by joint, Dave
turned it over to Sally—on the mission, he would have the job of un-
berthing and checking it, she would pick up the satellite and release
it, and he would berth the arm afterwards. Dave admired the way
Sally took charge of the arm. She was thoroughly familiar with it, of
course, from STS-7, when she had released and retrieved the SPAS,
and she was forever helping him out and telling him ways in which the
arm in orbit worked differently from the one here. For one thing, in
space it worked a lot more smoothly, never oscillating or jerking unless
the brakes (there was one for each joint) were deliberately jammed on.

Dave Leestma is a young-looking astronaut with a turned-up nose
and broad grin; he not only looks cheerful, he *is* cheerful—an instructor once told me that he has a "step-function" personality, which is engineering talk meaning the opposite of gradual change; Dave tends to
go from an ordinary mood to extreme enthusiasm almost instantly,
with nothing in between. (I never saw him go the other way, to an un-
enthusiastic mood.) Dave, who was then thirty-five, had been born and
brought up in Michigan, though his parents—his father is a minis-
ter—had moved to California, and he was graduated from high school
there. He is married to a Texan and has three children. Although
Sally was at the moment more proficient in slinging the arm about,
Jim Arbet, the instructor, told me Dave was learning very fast; he had
the kind of mind that was always on the alert, constantly asking what
would happen if he pushed this switch, then that one. Dave's reputation for brains at Annapolis was clearly based on a speculative, constantly questioning mind.

Sally articulated the arm, cocking its elbow far up in the air so
that the end effector could be brought inboard, directly over the pin on
the payload; it looked like the arm of a person trying to scratch his
shoulder. Because the end effector moved no faster than two inches a
second, the arm took forever to get anywhere, as if it were in slow motion. Jim Arbet watched Sally's eyes to make sure she was looking at
the right places—out the window at the pin, or at a couple of television
monitors above the console, to the right. The two monitors, whose
screens can be split into two images, throw up as many as four pictures
from cameras on the arm's wrist and elbow, and from cameras in the
four corners of the cargo bay; by their aid, it is possible to pick up a

payload without looking out the window at all—necessary if the grappling pin is out of sight, as would have been the case with the satellite, if Sally hadn't objected. Even with the pin well in sight, the wrist camera was useful, because coordination of the fifty-foot boom was difficult with the eye alone; by means of the camera, the end effector could be lined up with a target next to the pin, so that it could be driven home. When Sally had managed to fit it over the pin, she pressed a switch on her console which caused it to tighten around the pin, and then another button that pulled the end effector firmly against the satellite. Then she pressed more switches to release four latches that held the satellite in the bay and gradually lifted it out; by using the cameras at the corners and the split screens, she could observe all the latches at once and see that the operation went smoothly.

Visibility was not as good as it might have been because the payload—a mock-up of a giant cylindrical satellite called the LDEF, or long-duration exposure facility—took up almost all the compartment. It was a payload of Crippen's 41-C mission, now less than a month off; he would leave it in space, where hundreds of biological and chemical preparations would be exposed to vacuum and weightlessness for over a year, before the LDEF was retrieved and returned to earth. The balloon that was its mock-up looked like a small blimp (and, of course, had once acted like one by flying up to the ceiling). It would be several months yet before Sally and Dave had balloons of their own payloads; in the meantime they would have to practice with the LDEF both here and in the mission simulator. She and Dave both practiced grappling the LDEF and lifting it up to its release position, which was not much different from their own satellite's. (Each time they grappled, a technician on a catwalk along the cargo bay tethered the satellite to the arm, in case Sally or Dave accidentally let go of it.) The two astronauts had to know how to do each other's jobs on the mission, so that they would be able to back each other up. Every time they reached the release position and put on the brakes for the joints, the arm twanged. When the brakes were taken off, the arm twanged again. The same would happen with the arm in space, Arbet said. As the brakes had to be taken off before the satellite could be released, so that the arm could withdraw, all this twanging would impart to the satellite what Arbet called "tip-off rates," or a wobble. If they waited a couple of minutes, though, the vibrations would damp out, and there would be no tip-off rates. They in fact never released the LDEF; each time they raised it up, they brought it down again to practice docking it. Arbet was encouraging them to use the television monitors as much as possible, as it was good practice; they tried to bring the LDEF down so horizontally that all four latches would snap at once, something they never succeeded in doing.

At the end of the session, Dave berthed the arm. As she left, Sally said to Arbet, "Next time, no windows. Time after that, you'll fail the cameras. Time after that, no windows, fail the cameras on the dark side of the earth. Right?" Arbet told her she had the picture.

I saw them a couple of days later, in the mission simulators, where Jim Arbet told me they would be practicing the same exercises they had done in the manipulator-development facility. Though the facility had certain advantages, the simulators had others; among them, the controls in the crew stations were more realistic. Ted Browder and Shannon O'Roark were there; they would get the simulator running and then leave, because they, and the rest of the team, had no real roles to play in what was to be a simulation of the arm. They were having trouble, though—as sometimes happened, the simulator computers were acting up, and there were some glitches. "Hey, wait a minute!" Sally's voice came over the earphones from the crew station, where she had called up the graphics of the payload bay and the LDEF. "There's something missing. What is missing from this picture? No LDEF!" Indeed, the payload bay was there, but it was empty. Shannon, who was in charge of making sure the graphics were all right, said, "As far as I know, it's there, but you can't see it. Kind of hard to pick up if you can't see it." She twiddled some dials, and miraculously it appeared.

With the simulator up and running, Dave uncradled the arm and began trying to pick up the LDEF. Sally flew the orbiter from its rear controls, but there wasn't much for her to do; once, when Dave asked her to move the spacecraft so that he could get the LDEF to the proper launch position, she yawned. "Thought you'd never ask," she said. As the arm slowly inched the satellite upward, she declared that the operation was a real "watch-the-grass-grow" exercise. She abandoned ship and appeared shortly in the instructors' room, with a cup of coffee.

Even Dave was getting bored; he was playing with the switches for the lights in the cargo bay.

"It looks psychedelic," Jim Arbet said.

"Looks like a light show," Sally said. "There ought to be music."

Dave at last got the LDEF up to its deployment attitude and released it. He asked Sally to come back to the crew station; it was her turn to run the arm. Dave appeared in the instructor's room, with a can of Diet Coke. He told Arbet that the simulator and its graphics gave him better training in the sort of hand-eye coordination that he would need to fly the arm in space than the manipulator facility had. "The arm is as close to flying as I'll get, except for flying in the back seat of a T-38," he said. He was hungry; it was almost four in the afternoon, and he hadn't had anything to eat since ten that morning, when he and the rest of the crew made a meal over the one-G trainer, using the same type of food they would have aboard, and then eaten it—they had had unsimulated Spanish eggs, unsimulated sausages, and unsimulated granola, rib-sticking fare, but that had been six hours ago. The class had concluded with a lesson in stowing unsimulated trash.

At length, Sally told Dave to come back and take over. When he got back to the crew station and looked at the visuals, he couldn't find the arm. "What have you done with it?" he asked, alarmed. He turned on the arm's wrist camera and saw the earth appear on the visuals. "Oh, I see—you put it over the side, and underneath." There was a peal of laughter from Sally as Dave began the watch-the-grass-grow procedure of getting the arm back inboard and berthing it. It was the sort of problem Shannon might have thrown into a sim—she and Sally were not unalike. It was, of course, the sort of mischief Sally enjoyed, like bringing up the SPAS on another occasion when she was bored. Clearly, training had its dull moments the second time around. I also thought that Dave, by turning on the wrist camera, had recovered the situation nicely.

Surfing into Honolulu Airport

The week just before Easter, which occurred late in April, was the week after Crippen's 41-C mission returned. In addition to launching the LDEF, those astronauts had succeeded in intercepting the Solar Max satellite, which had been in orbit for three years but which had been out of commission most of that time, capturing it with the arm, replacing the defective parts, and returning it to orbit. One of the crew, George Nelson, had flown out to it on a jet-powered backpack called the manned maneuvering unit, docked with it, and tried braking it with his jets; however, he had been unable to dock because of an obstruction, and in the course of his attempts, he had caused the satellite to pitch and yaw as well as roll. As a result, Sally Ride and other members of Crippen's old STS-7 crew, who had had experience with the arm when they had released and recaptured the SPAS, had been summoned posthaste to the fixed-base mission simulator to see if they could determine whether the wildly wobbling satellite could be captured; Sally found she was just able to do it. (The technique, it turned out, wasn't needed, as ground controllers in the Solar Max's remote control center at Goddard had been able to stabilize the satellite.) "I had to thank Crip for giving us an STS-7 reunion," Sally told me— around NASA, apparently, old crews never die, even though the members go their separate ways. "The only person who wasn't there was Crip himself." On a second EVA, Nelson and his partner, James D. Van Hoften, repaired the satellite, which later was put back in orbit by the arm.

The air-to-ground commentary was piped all over the Johnson Space Center, and the members of the 41-G crew had paid close attention whenever they could—"Not only because of Crip, but because we all had a lot of friends aboard the mission," Sally said. This, of course, would be true to a greater or lesser extent of all the missions, but no one pays closer attention to a mission in orbit than the crews that are following along. For one reason, every mission that is flown brings them one mission closer to launch. With 41-C out of the way, there were just two more flights ahead of 41-G—41-D and -F. "So, it's as if we were on the runway, third in line for take-off," Myron Fullmer, the crew's training supervisor, told me. As a crew moves ahead in the line, coming closer to take-off, they find that the pace quickens and that people pay more attention to them. They have a higher priority in the simulators; crews nearer launch are always able to bump crews in back of them if they are behind in training. 41-G was now one place higher in the pecking order. (Highest in the pecking order, of course, is the crew that is aloft; for STS-7's reunion in the interests of 41-C's capture of the Solar Max, 41-G lost a four-hour session in the mission simulators.) Upcoming crews also pay attention to ongoing flights to see what they can learn. "It's as close to being in orbit as I'll get until I go up," Leestma told me. "Of course, I listened carefully because 41-C had an EVA, and I'll be doing an EVA; but I'd listen anyway, even for routine stuff. I get a feel for how the timeline flows—whether the crew sounds relaxed or not. I get a feel for the best way to talk to the ground. When a crew comes back, it always debriefs the entire corps. There really is a good mechanism for passing on all this experience."

Space flights are not isolated events, as they appear in the press, but parts of a cohesive fabric, a continuum, not only with respect to their origin in NASA's manifest but also with respect to their training and execution. The Johnson Space Center is a spawning ground of many missions, with the same group of people essentially producing all of them—even the astronauts, who, though they train for and fly individual missions, are part of an astronaut corps whose collective job it is to fly all of them. The digesting of experience from past missions, the planning of future ones, and the flying of current ones go on simultaneously.

The crew, of course, had seen Crippen almost as soon as he got back: they were all out at Ellington Air Force Base, about ten miles from the Space Center, where he and the rest of the 41-C astronauts had flown in, in one of NASA's Gulfstream jets, from Edwards Air Force Base, where their orbiter, the *Challenger*, had landed. McBride had talked with him the morning I arrived; among other things, McBride had shown Crippen the 41-G crew's emblem, which had just been drawn by an artist in NASA's graphics department, with heavy advice from the crew. Most mission emblems are fraught with symbol-

ism, and this one was no exception. In the middle was the gold pin all
astronauts receive after they first go into space: three golden rocket
plumes ending in a golden star, the plumes held by a golden circle representing an orbit. The upper half of the emblem, a big circle, contained a flapping American flag; the lower half was the sky with some
of the constellations and stars the astronauts would use as guides, including the Pleiades that McBride had had so much trouble finding
prominent among them. One cluster of five stars symbolized the five
astronauts. Curving along the bottom were their names with the appropriate male and female symbol after each, for with only two women
aboard they were proud of being the first truly mixed crew—Sally
Ride's presence on the STS-7 mission, it could be argued in a place that
loves to find the most recent event is a "first," had been an anomaly, a
form of tokenism, in an otherwise all-male crew. Crippen approved the
emblem. With Crippen back and the emblem in hand, the crew felt
that things were beginning to fall into place.

Though Crippen was back, and the crew had seen him around and
about, he had not yet become involved in their training; there are so
many debriefings and press appearances after a mission that it would
be two or three weeks before Crippen was free to join them. He was, of
course, making his opinions known, as he had done before 41-C. I overheard a heated discussion between Myron Fullmer and Browder, in
Fullmer's office, where Fullmer was informing Browder that Crippen,
for his first sim with the 41-G crew, wanted to do an integrated sim on
May 8. Integrated sims, which involve all the flight controllers as well
as the astronauts, are principally done by crews who are in the last
month or two of their training. If such a crew is unable to be present
for a scheduled integrated sim of its own, its place is offered to a crew
farther back in line. Because the purpose of integrated sims is to train
the flight controllers at least as much as the astronauts, there is reason to go ahead even if the prime crew can't be present. To fill in for a
senior crew is regarded as a challenge—an honor, even—to the crew
that takes its place, and Browder was not sure that 41-G was ready,
particularly as McBride, Ride, Sullivan, and Leestma would have had
hardly any time at all to practice with Crippen. Browder pulled a long
face when Fullmer told him that Crippen thought an integrated sim
was a good way for him to see how the crew stood, and to get himself in
tune with the crew. Browder didn't like the idea; an integrated sim
was a very public performance, where the astronauts would be under
the eye not only of all the flight controllers but also of a large new team

of instructors whose scenarios might be very tough. Browder said he didn't want his crew to fall on their faces—or, as he put it, in a metaphor that was something less than apt, "I don't want them to fall on their swords."

"Maybe, if they make a hole in themselves, it would be a good thing; it would show them that they need more training," Fullmer said.

"If you fall on your sword, you don't want to make a hole in your heart; better in your arm," Browder said, giving the metaphor a good stretch. "It'd be better for Crip to try the crew first in a stand-alone sim, with nobody watching."

Fullmer said, "Perhaps they won't fall on their swords. And besides, it's what Crip wants."

Browder had no choice except to give in. He told me later that even though the purpose of all the training was to get the astronauts ready for a flight, he himself couldn't help but feel that the immediate purpose of all the stand-alones was to prepare the astronauts for the integrated sims, and consequently he didn't want to put them to that test prematurely. When Ted took the matter up with Crippen later, Crippen agreed with him, but he still went ahead with the integrated sim.

I overheard another conversation, this time in Browder's office in Building 4, between him and Randy Barckholtz, his orbiter systems instructor. Randy remarked that without Crip, the crew had no real *commander*—despite the fact that the role was filled in part by Sally, who did the most to relay Crippen's wishes, or McBride, who as pilot was number two in command, or by whoever happened to be sitting in the commander's seat during a sim, Leestma or Sullivan. Browder was delighted that McBride—who believes command in a spacecraft is looser than in a navy vessel—was not trying to be commander. As any rookie pilot does, McBride—who was spending most of his time learning—had the farthest to go, and he was getting good piloting experience in flying ascents and descents in the simulator that he would never get if Crippen were there.

After Randy left, I asked Browder whether he thought the crew had missed anything in their training as a result of Crippen's absence. "I'm sure my managers will be asking me the same question—how training goes with the commander absent for the first half of it, which will probably be the wave of the future," he said. On balance, he thought it worked out very well. He hoped to have the crew ready for Crippen when he joined it. The crew was by no means a headless group crying out for a leader; in addition to perfecting their individual skills, they were beginning to work well together. And in many ways it was better that McBride did not function as a commander, because if he did, patterns might develop that would make it more difficult for Crippen, and the rest of the crew, when Crippen returned. McBride himself said to me, "There would be a lot of things we'd have to unlearn. We try to keep things flexible for Crippen."

With respect to Crippen himself, he of course had missed nothing by being absent during the generic part of the training—the basic training, applicable to all missions, that any untrained crew has to start off mastering. Crippen knew it all. What he had missed, undeniably, was getting to know this particular crew and how they interact. And the crew didn't know him, either, at least in his role of commander, no matter how often Sally Ride told them his particular preferences, such as that he always liked to do star alignments himself. Of course, the crew wouldn't *be* a crew until Crippen arrived and took his place at its head, and in this respect it was behind where it should have been. "I expect he'll catch up with this in three or four weeks," Browder said. "If not, I'll raise a flag." The crew was even more optimistic than Ted. "We know Crip," Kathy Sullivan said to me. "We haven't been cut off from him. And of course Sally knows his way of doing things. He's been on several other crews, so he'll know which reins to grab. He'll crystallize our organization."

Sullivan's mention of Crippen "crystallizing our organization" reminded me of a conversation I had had earlier with John Young, and of what was really at stake here. I had asked Young what NASA was basically trying to achieve in crew training. "We want the crewmen and -women to stop working as individuals and work together as a team," he said. "This happens after about a thousand hours of training—usually a month or two before launch, when the crew is doing integrated sims. I listen to the communication loops, and I can tell when they're there. Suddenly the team sort of crystallizes. All of a sudden a crew of three can do the work of a crew of four or five that hasn't crystallized. When that happens, they solve problems automatically, without much discussion. When it happens, it's like the difference between night and day. And Crip on 41-G is a sort of test case. We'll see how far down the line we can go with the commander absent, and still have the crew come together before launch."

When I caught up with the crew (still minus Crippen) and the team in the mission simulators, I found they were all in high spirits. Part of it was that it was the week before Easter, and they were ribbing each other about having to work on Good Friday. (NASA did not give Good Friday off, but some of the contractors who paid some of the instructor's salaries, such as Rockwell International, who technically employed Ted and Randy, did. They had to come in anyway.)

McBride said that normally by now he had all his Easter eggs dyed.

"Jon's getting ready for Easter weekend," Sally said. "Any rabbits come around his house, he'll kill 'em."

"Any rabbits come to Jon's house won't find him home," Ted said.

I wasn't sure whether these high spirits were a sign that the crew was crystallizing or not. I didn't think it was; there was altogether too much chatter for an efficient crew. Still, on some level, it probably helped the process along.

Inexplicably, in ground calls, the crew was calling the team—in its role of Mission Control—by such names as Peoria, Schenectady, and Poughkeepsie, instead of Houston; and the team was calling the crew by names like "Intrepid" and "Wasp" and even "Potemkin" instead of "Columbia," the name of the orbiter they were supposed to be flying. When I asked Browder about this, he explained that they were all making fun of the fact that almost everything about the mission had come unstuck: NASA was looking into substituting the *Challenger* for the *Columbia,* which was still being renovated at Palmdale; NASA wanted to keep it there in order to remove its bulky ejection seats, which had been needed for the earliest flights, and which would require another six months to take out. They took up so much room that they greatly reduced the space available for the crew, and it turned out that NASA was thinking of adding to 41-G one, and perhaps even two, additional passengers—payload specialists, non-NASA scientists or technicians with particular experiments to perform. The *Challenger* would be roomier.

In one way, the 41-G crew preferred the *Columbia,* even though it was older and more cramped. They could stay in orbit longer with it (ten days instead of eight) because the *Columbia* had been fitted with extra oxygen and hydrogen tanks for supplying the Spacelab, the laboratory module it had carried in its cargo bay on the ninth mission. A drawback with the *Columbia,* though, was that it could not come back

to the Kennedy Space Center, the most desirable place for orbiters to
return because they can be serviced more quickly for their next
launch: they don't have to be flown atop a Boeing 747 all the way across
the country, as is the case when they land at Edwards Air Force Base
in California. The *Columbia* lacks a critical navigation aid called a
heads-up display, which flashes flight information onto the windshield
so that the commander and pilot do not have to take their eyes off the
runway—essential for landing at Kennedy, with its likelihood of
crosswinds and its landing strip hemmed in by swamps and rivers full
of alligators; Edwards, with several runways laid out in different directions on a broad, dry lake bed, is more forgiving. The *Challenger*
has the heads-up display. Even with the display, there was no guarantee of landing at Kennedy; indeed, twice, on STS-7 and just now on
41-C, Crippen, flying the *Challenger*, had been waved off from Kennedy due to bad weather and had to land at Edwards instead. All else
being equal, he would like another crack at it, and this he would not
get in the *Columbia*. If there was a switch in spacecraft, even the date
of the launch, now scheduled for September 1, would have to be postponed a week or two because the *Challenger* was in the workshop at
Kennedy having several hundred of its thermal tiles reattached more
firmly. (Since 41-D and -F would be using the recently delivered *Discovery*, the third in NASA's fleet of orbiters, those missions would not
be affected by whatever was decided about the *Columbia* and the
Challenger.) With so many major issues—launch date, landing site,
spacecraft, and number of people aboard—all coming unstuck at once,
the astronauts felt as though they were coming unglued themselves.
For the time being, Holmberg, the manager of their crew activity plan,
sensibly shelved the 41-G flight plan and turned his attention to another mission assigned to him. The astronauts, however, had no such
easy out. "So you can see why, with everything else breaking loose,
they might think that Mission Control might be almost anywhere,"
Browder said.

That day they were practicing orbital skills—what they would be
doing in orbit, or in what Browder and an increasingly large number
of people call "on orbit." (This misuse of the word "on" in "on orbit"
probably hit the Space Center at about the same time as its misuse in
the phrase "on line" hit the rest of the country. "On line" is where you
bring a piece of machinery when it's ready for action, whether it be a
generator, a piece or artillery, or an inertial-measurement unit. One
cannot get *on* an orbit. One stands "in line," and one flies "in orbit.")

When I arrived, Randy Barckholtz was putting a bias on a sensor in an oxygen tank that fed one of the fuel cells, where oxygen and hydrogen were combined to generate electricity. "Biasing" a sensor means causing it to give a false reading. Sensor problems, especially in fuel tanks but also in the electrical system and elsewhere, were a special class of problem that the team used often, because they had been experienced on many flights and because it took a minimum of effort with the light pen to create a maximum amount of havoc with the crew, who had to decide whether or not a wild reading on a pressure or temperature gauge reflected a real problem. The team was apt to throw in sensor problems whenever things got dull; among nits, they were king. As a result, an amber caution-and-warning light flashed in front of McBride. All four crew members turned their attention to the problem, advising McBride, who was responsible for the fuel cells, to check and see if there was a drop in the voltage being fed to the particular bus (a distribution panel) that was being fed by the cell that derived its oxygen from the tank in question; possibly the tank was leaking.

"I don't like them all working on the same problem," Browder said. They were still in the committee stage. He turned to Shannon O'Roark and asked her to throw in a bite. A bite is a nit or minor malf, and the one he wanted was for the computer data screen the astronauts were looking at to read the pressure sensor. It went blank. Kathy Sullivan and Dave Leestma instantly set out to solve that problem, while Sally Ride stayed with Jon McBride as he continued to work on the oxygen pressure problem. They were much better than they used to be, Ted said; at least they no longer just stood and looked at each other when an alarm went off.

He seemed pleased, not only with the astronauts but with himself. As Team Lead, he had to do a fair amount of orchestrating—he had to decide when the pace needed to be sped up or slowed down, when the team's killer instincts had to be restrained, and when a crew had to be thrown a problem to make two astronauts get out of the hair of the other two. And as the scenarios, and the crew's capabilities, got more advanced, there was more orchestrating to be done. He was particularly interested in getting the crew to learn to set their priorities so that, when there were several problems at one time, they could decide which ones should be tackled first, and then divide up the work between them. "They have to learn to do that themselves, so that prioritizing will be instinctive," he said. "If Crippen had been here all along, he would have done that for them. So in a sense, it's an advantage to not have him here." Setting priorities was clearly an important skill to master before a crew crystallized.

The computer screen problem had to be fixed before the oxygen pressure problem could be resolved. At length it was back in business. "I hope you noticed the wonderful thing I did," Kathy Sullivan said, and Shannon and Ted warmly congratulated her. Of the crew, Kathy was perhaps the least experienced in orbiter systems problems; Jon and Dave had extensive backgrounds flying navy aircraft, and Sally of course had been through the training before, but Kathy was a geologist with no flight experience at all. Though she was the same age as Sally and had much the same background—she had grown up in California and had a doctorate (from Dalhousie University in Canada, awarded in 1978)—she was quite different. She was quieter, less glittery. Her blue jeans were not Calvin Kleins but from a more serviceable emporium. She was bigger boned than Sally and had a broader face, and altogether looked as though she would be happier outdoors. She was very industrious and worked hard to learn more about orbiter systems, often signing up for extra lessons in the single-system trainer to bone up on things she wanted to find out about, even though, as a mission specialist, she had no real obligation to do so. When I attended a session in the trainer with her once, she told me, "I want to get more familiar with the cockpit—the more I know about it, the more at home I'll be. Besides, I never was one to be extra baggage." Kathy seemed to dig a little more deeply than Sally, who, with all her brightness, seemed to skim lightly over the surface. In the single-system trainer, Kathy was obviously concentrating hard—astronauts generally seem to have a power of concentration greater than most people's, a quality that may stem from their compelling interest in what they are doing. She told me, "I find training exhilarating. I rarely feel burnt out. I don't want to focus so hard I miss something else, but when you get into a simulator and start working on a problem, there surely is a dissociation from the rest of the world." She was peppering the instructor with questions, not all of which he could answer. "In the single-system trainer, I come out with any question I've got, even my dumbest questions, to solicit more information," she said. "You can't do that in the mission simulators. Here, we can freeze the model any time we want and talk about it, in a way you can't when there are four other guys with you. You have the liberty to pause and come to a fuller understanding." In the mission simulators, Ted and the team gave Kathy all the encouragement they could; they seemed to enjoy helping her along. At present, though, Ted told me, Shannon did not have to dig very deeply into her bag of tricks to occupy the crew while either Kathy or Dave was standing in for Crippen as the computer expert.

With the screen back on line, Jon quickly found out that the oxygen pressure leak was no leak but a sensor problem, and he reset the limits on the sensor so that it could no longer go out of bounds, setting off an alarm.

No sooner had he solved the oxygen pressure problem than Andy Foster, on the Control console, threw at him a leak in a manifold, in this case a chamber that fuel entered from one pipe but left from several, supplying several thruster jets. Andy had arrived on the team only a few weeks before, replacing Bill Russell as Control instructor, in charge of the various propulsion systems and the aerocontrol surfaces such as the rudder, ailerons, elevons, and speed brakes. He was a quiet, even-tempered man who, alone on the team, never seems to get gremlin-like, mischievous pleasure out of screwing up the astronauts. Perhaps this was because he was new at the game. He introduced nits into the system with a reluctant, almost sorrowful air, as though he didn't entirely approve of dirty tricks; his long, reflective face often looked sad when an astronaut fell into a trap. He is a methodical, straightforward teacher—he had been in the navy for twelve years, where he had been a flight engineer like Leestma and also a flight instructor, so he took readily to teaching astronauts.

The manifold leak did not show up as a leak in the thruster system's fuel (the reaction-control system), but rather in the orbital maneuvering system, because in the rear of the spacecraft, where the problem arose, the thrusters, as I had learned, could draw fuel by a cross-feed arrangement from the maneuvering system tanks, and it was doing so now. Hence, in trying to track down the leak by isolating various tanks and pipes, in series, closing isolation valves one after another—as I had seen him do in the single-system trainer— McBride followed all kinds of wrong trails.

"Come on, Jon!" Andy, who always seemed to be plugging for the astronauts, said.

"Jon has misinterpreted some clues—he's going down the primrose path!" Ted said, if not with glee, then with something close to it.

Andy did some swift work with his light pen and then informed Ted that he had slowed the leak down in order to give Jon more time to solve it. Browder nodded in approval; he was usually the one that was slowing down the other members of the team, and he was clearly in agreement now. Like many a rookie pilot, Jon was still having difficulty mastering the thruster system—it is one of the most complex systems, with all sorts of redundancy built in. This situation, among many others, was one of the reasons Ted was enthusiastically in favor of the system NASA was now employing of keeping one team of instructors with the crew of astronauts all the way through training, among other things allowing them to follow up on any particular astronaut's special requirements. Up through the time of the third or fourth shuttle mission, the teams and crews were randomly assigned to the simulators, on a sort of catch-as-catch-can basis, for each session. This shotgun system, as Browder called it, had its advantages; each team got to see a variety of crews with their strengths and weaknesses, and each crew was exposed to a variety of instructors. The great disadvantage, of course, was that the frequent switches in instructors meant that there was no continuity in the training; problems that came up in one session could rarely be followed up in another.

The new system, however, allowed a more structured approach. This extended all the way from scripting, when a team would aim a problem at a crew's known weak points, to repeating problems that a crew had messed up on earlier sims, to devoting special attention to an astronaut's particular needs, as Ted and Andy were doing now. "Having a crew dedicated to a team adds a whole new dimension to training," Ted told me. "The old way, we never got to know the crews so well. We see them on a daily basis now. We see them personally—we go out for beers together on Friday evenings. The crews always remember their roots—they are always the first to share their accomplishments with their trainers." The teams obviously reciprocated the loyalty and affection of the crew, as I found out later at the launch, when the team was so overcome. Training in this fashion was obviously a very intense experience for all hands. "It's very close," Ted told me once. The crew and the team clearly identified with each other. "I've seen some of the astronauts I've worked with pick things up from the instructors, even adopting some of our characteristics—it almost allows us to fly in their bodies. And after a mission when they get a pat on the back for a job well done, we feel a little of that. And the same for their errors." This held true even in the eyes of the rest of NASA: if a crew did especially well on a mission, it would redound to the glory of the team that trained them, and the reverse was true as well. The crew and the team were almost alter egos.

The main drawback of having one team stay with one crew was if the team had a weakness—say, it wasn't as proficient as it might be in one system or another. Then that weakness would be passed on to the crew. However, such deficiencies, if they existed, were apt to show up during integrated training, when a different set of instructors took over, and it could be corrected then. NASA had another way of overcoming the somewhat ingrown nature of one-crew–one-team training: unlike an astronaut crew, who shared the same office on the third floor of Building 4, the instructors were not assigned to sections by teams but by positions—that is, the Control instructors were in one section, the Systems people in another, and so forth. That way, the instructors could all keep up with what the other teams were doing, and the teams could generally cross-fertilize each other.

While Jon was still trying to solve the leaking thruster system problem, Randy threw him what appeared to be another leak, this time in the hydrazine fuel used to power the motors that drove the pumps for the hydraulic system, the auxiliary power units, used during entry. Jon solved this problem in a jiffy: there were two sensors that gave readings for the same tanks; since one was giving a danger signal and the other wasn't, he deduced correctly that the danger signal was a mistake. Then he turned back to the thruster system leak. When he still couldn't solve that one, he took what was an easy way out—instead of finding and isolating the leaking manifold, he turned off the entire thruster system and switched to the redundant system, the verniers. There are many fewer vernier engines—six over the entire orbiter, as opposed to thirty-eight for the main thruster system. In addition, the vernier jets are much smaller: the main thrusters each put out nine hundred pounds of push in a flame that shoots forty feet, and they can, as an astronaut once told me, "really horse the orbiter around." The verniers put out only twenty-five pounds each, enough for fine-tuning attitude. In this sense, the two systems are not exactly redundant but complementary, as are the major and minor adjustment knobs on a telescope; using the verniers now was like swinging a telescope from one horizon to the other with the fine-adjustment knob—it would take longer.

Browder looked unhappy at McBride's solution, not only because it expended more fuel and was not as elegant as finding and isolating the particular leak (a nice problem, because several thrusters would have to be turned off, forcing Jon to find and turn off their opposite numbers in order to balance the system), but more importantly because Andy had a major malf coming up that required the astronauts to be using the thrusters. "We'll have to get them back to the thrusters, by giving them a bigger leak in the verniers," Ted said. This was, he said, another example of how a sim tended to have a life of its own, apart from any scenario—sometimes the astronauts do unexpected things that throw a monkey wrench into the best-laid scripts, and the team always had to be ready to improvise.

A little later, when he made a ground call to McBride to apprise him of the new leak in the verniers, he consulted with the rest of the team to decide in what city Mission Control would be located this time; they selected Moundsville, which is where the state prison is in McBride's home state. Conceivably they chose it because that, at the moment, is where they wanted to put Jon, for upsetting their plans.

"They're in West Virginia!" McBride remarked with pride in his voice, when the call came through.

"Someone down there must have an atlas," Kathy Sullivan said, giving the impression that no ordinary mortal would be expected to know the name of a town in that state.

"We're overlooking the beautiful West Virginia State Prison," Randy Barckholtz said.

"A lot of people would like to overlook West Virginia," Kathy said. Now that she was solving problems in the simulators and beginning to feel more at home there, she proved to be as good at making one-liners as Sally. In any organization where there are a lot of young people from all over the country—such as the armed forces and NASA—jokes about home states and home towns abound, and they are endemic to the natural banter of astronauts and instructors. A little later, when the astronauts were calling the ground, they decided that Mission Control had moved—"Where do you think they are now?" McBride asked his crewmate, in a hurried conference. They picked Tippe City, the Ohio town where Shannon came from. As it happened, Shannon had just killed all three inertial-measurement units, so that the astronauts had to spend the rest of the session regaining their alignment with the crew optical alignment sight. They had troubles. First of all, they had difficulty seeing the stars because the visuals were suffused in an orange-red glow.

"I didn't know space was orange," said McBride, who was doing the alignment.

"It's probably all the orange-mango juice we've been drinking," Sally said. Orbital sessions are slower paced than ascents and descents, and all the astronauts had been raiding the cooler below decks, which was filled with the same snack foods they would have in flight. They had all been making heavy inroads into a newly-developed space drink.

"As we come into the dawn, we've got the Pleiades," Kathy was saying.

There was a bump in the spacecraft as McBride banged his knee against a console while he was trying to look for the alignment tele-scope. As he was one of the biggest astronauts in the business, he sometimes had a hard time getting around the simulator—floating around inside the orbiter in space would be much easier. "I've knocked off a scab," he complained.

"Shall I get the med kit?" Kathy asked. She came back with a small piece of cloth, which she said was "not going to be real sani-tary."

Sally Ride, having jumped ship, appeared in the instructors' room, wearing white slacks and a blue plaid shirt; she carried a con-tainer of orange-mango. She wanted to tell Shannon that some of the checklists for making entries into the computer for bringing up dif-ferent programs were wrong. "Sometimes the ones for the orbiter are different from the ones in the simulator, but these won't work in either place," she said. Sally watched on Shannon's visuals as the stars shot across the red sky. Evidently Jon was having trouble with the align-ment. "Will Dr. Ride please come and help us find some stars?" he said over the loop, and Sally disappeared.

"There's Sagittarius," Sally said at length, back in the crew station.

"If that's Sagittarius, I'll eat it," Shannon said.

"There's Orion," McBride said a little later.

"I don't see any stars that belong with it. I see the belt but not the knife," Kathy said.

Shannon explained that that was because they were not looking at Orion, but at Hercules.

"Hey," said Jon a few minutes later, "Do you really want us to do this alignment with no lights and no power for the telescope?" Randy Barckholtz had popped a circuit breaker.

Everyone, I thought, was having a very good time. Regardless of whether the crew was any closer to crystallizing, the astronauts and the instructors seemed to be becoming more and more of a unit. They were playing off each other, like the performers in a vaudeville act. I wondered if the good spirits and general bonhomie would survive the arrival of Crippen.

*T*he next morning, following their orbital skills session in the fixed-base simulator, the crew was in the motion-based simulator at 8:30 for four hours of emergency de-orbits—hasty returns to earth if something goes wrong in space so that the mission must be aborted. They seemed just as cheerful as they had the day before.

"Looks like it was a bad day in orbit for somebody," Sally said as she looked over the control panel, which spelled trouble for the previous crew.

"Let's eat," said Kathy. Alone on the crew she was unmarried, and therefore was the least likely to have had breakfast. They simulated eating a space breakfast, but they found little of substance in the food locker.

"What have we got?" Leestma asked, rummaging around.

"Orange-mango! Strawberry drink!" Sally reported.

While they drank their breakfast, McBride was telling the rest of the crew that over the weekend he had seen the Blue Angels fly in El Paso. He still had a number of friends among them, left over from his own barnstorming days with the red, white, and blue airplane.

Shannon, who was listening in, said she wanted to go and have a ride in one of their planes.

McBride said he would take her over, and after that she was on her own to cadge a ride.

While all of this was going on, the team was resetting the switches on their consoles for orbit once more (all the descents would start from there) and looking over the scenarios for the morning.

"We'll need the cabin leak right away," Ted said to Randy, who was in charge of such things.

Randy, looking at the script, told him, "It says here a six-inch hole in the cabin! That would be big enough to pull people through! It must mean six-tenths of an inch."

The astronauts, too, were resetting their switches, talking non-stop as they did so.

"Have you been to Poughkeepsie?" Sally asked.

"What's that? An island?" Jon asked.

"It's a town where IBM is," said Dave.

"Have you been to Moundsville?" Sally asked.

"Yes," said Jon. "And there's still an Indian mound there."

"That's about all there is," said Sally.

"They remember yesterday," Browder said to his team.

"They're sounding a bit frisky," said Randy. "We'll have to settle them down."

"Are you ready?" Ted said to the astronauts.

"We're all up and around," said Dave. The crew made waking-up noises, as if they had just responded to the morning call.

"Not quite ready," said Jon. "I'm going down for another orange-mango."

"Bring me a strawberry," said Sally.

"It's getting boring up here," Kathy said.

An alarm clanged in the spacecraft.

"Oh, dear!"

"Oh, my goodness!"

"Goodness gracious me!" (In space, astronauts are encouraged not to use four-letter words, which may be one reason why, in the simulators, the astronauts sometimes practiced with epithets whose purity was unquestioned, and which their great-grandmothers might have used. It may also have been a polite accommodation to the fact that there was a mixed crew—indeed, when something seriously bothered anyone, they were apt to come out with a polite "Oh, rats!")

"It's a cabin leak!" said Dave.

"What is the cabin pressure?" Jon asked.

"It's down to 14.2 pounds per square inch, from 14.7," Kathy told him. The spacecraft atmosphere had the same pressure as the earth's at sea level and the same ratio of the two major gases, approximately 80 percent nitrogen and 20 percent oxygen.

"We're in trouble," said Jon.

"Good call!" said Sally. They all laughed.

"Let's look for the leak," said Dave.

Randy said to me, "They will look for it, but they won't find it." Rarely is a leak, of whatever size, anything as simple as a puncture, possibly caused by a meteor; more often it is caused by a leaky valve, or a leaky seal on a hatch or a window, or a valve that was open that should be shut. Dave had pulled out a checklist of things to look for, but the cause wasn't on it. Randy himself didn't even know what the cause was—it was simply a leak—because the object wasn't to have them find it and fix it, but to force them to come back to earth quickly.

Dave and Jon had shut everything it was possible to shut, but none of it did any good. They then took out the emergency de-orbit checklist, which was a version of the normal de-orbit checklist, stripped to the bare essentials, with many lesser items left out. They all knew what they had to do: it was Dave's job to shut the payload-bay doors. Kathy Sullivan closed out the middeck area, making sure all cabinet doors were shut, everything was secured, and the seats for herself and Dave were set up. Meanwhile, upstairs, Jon and Sally were figuring out how much time they had before the leak reached a critical point; they could survive and be all right if the pressure inside the cabin didn't drop below four pounds per square inch. (The fact that there was so much nitrogen in the atmosphere meant that there would be a danger of getting the bends.) Before they reached that point, they would fire into the cabin every bit of gas they had. With a hand calculator they figured that they had just under two hours, by which time they had to be on the ground, or deep enough in the earth's atmosphere so that the leak no longer mattered. As it took about forty-five minutes to reach the ground after the de-orbit burn, they had perhaps an hour before they had to do it. Jon and Sally turned once more to the hand calculator to figure which of the dozen landing sites around the world where NASA had emergency arrangements would allow them to do a burn within an hour. "De-orbit for Honolulu!" Jon announced. Randy nodded his agreement—ordinarily, the ground would contribute to these calculations, but the purpose now was to train the astronauts, in case the situation arose when they were out of communication. He liked the way the astronauts were dividing up the jobs.

Because virtually everything astronauts do involves following checklists—whether it be simply completing a nominal launch or landing, getting through an emergency, tracking down a leak, or determining whether an offscale reading on a sensor is real—I once asked Kathy Sullivan whether she felt a slave to them, whether they left room for independent thought or action. She smiled. "There's a saying here, that there are two ways to screw up a procedure—to do it the way the checklist says, or not to do it the way the checklist says. Checklists are not necessarily a guarantee of success. They *do* make sure you don't forget things that are important, and in tracking down problems, they suggest avenues you should follow. In a system like the orbiter, there are many complexities. There are so many funny little system things—little small gotchas or hazards. There are far too many for any one person to know. And a checklist is simply a distillation, a focus point, of everyone's collective wisdom." Kathy was perhaps the most reflective of the four astronauts, the most thoughtful about what her training and going into space meant to her.

After about fifteen minutes, the cabin pressure was down to a little over ten pounds per square inch. Randy slowed the leak, to give them more of a chance—though lest anyone think he was being soft on the crew, he explained that diminishing the rate was authentic because, as the cabin pressure dropped, the leak would slow down. He said, "We won't give them any more anomalies, because we want to test their ability to do time-compressed procedures."

Ted said, "It's O.K. to give them some malfs, providing they're not too major." Ted not only had to know when to put the brakes on his team, but when to step on the gas. He told me he thought the astronauts were doing very well; they were a little ahead of the game, and so it was time for a curve ball or two. Even though they were still at nine pounds per square inch, Jon was getting set up to take certain actions the checklist said he should do at eight pounds per square inch, such as increasing the flow of oxygen into the cabin, cross-feeding from the supply designed for the fuel cells.

Ted pointed out to Shannon that the astronauts were making a mistake; they should be using the relay satellite for communications with Mission Control, but instead they were using the ground stations.

"Tsk, tsk," said Shannon. Communications were in her area.

"You want to give them something, since you didn't get a chance yesterday?"

"I'd love to fail a general purpose computer," she said eagerly. Ted said that failing one of the four on-line computers at this point might be a bit too much. He told her to give them a little nit.

Shannon struck with her light pen. "I failed a failure-identification light, part of the caution-and-warning system," she said. "They probably won't even notice it." They didn't.

I said, "Maybe you should give them something worse."

"Don't make her worse," Ted said to me. "She doesn't need help."

Shannon struck again. An alarm went off inside the spacecraft. "I killed a computer that will force them to do a one-engine burn," she said. There were, of course, two maneuvering system engines, one on each side of the tail; the computer she had killed governed the right-hand engine. Although in theory each of the computers could control the entire orbiter and all its systems from launch to landing, in practice not only were the different phases of the flight assigned to different computers, but so of course were the different spacecraft systems; moreover, different parts of the same system were often divided up and attached to different computers, in what were called "strings." For instance, the right maneuvering system engine might be on the same string as a quarter of the thrusters, one of the fuel cells, and half the environmental system. The other maneuvering system engine would be on another string leading to a different computer, along with other bits of equipment. This arrangement keeps all the computers busy and provides redundancy. When a computer dies, the strings of equipment it was commanding can be moved to other computers—a process called "restringing"—and Sally suggested to Jon that they set about doing this now.

"No, no, you don't have to do that, Sally," said Shannon, not so she could hear her. "I want you to do a one-engine burn!" She struck with her light pen again, killing a second computer, and thus making the restringing more difficult.

"Maybe we should do a one-engine burn," McBride said.

"Hooray! That's right!" said Shannon, again so that they couldn't hear.

"We're going to have to check the gimbals for steering, and we're going to have to turn off the good left engine manually, when it's fired long enough," Jon said.

"No, no!" said Ted. "They don't have to do that! When the engine automatically shuts down, they'll see they don't have to." When a computer fails and a maneuvering system engine can't fire the de-orbit burn, the computer with the good engine strung to it automatically takes over the problem of refiguring the length of the burn; one engine would have to fire twice as long to develop the right amount of thrust, and when that is accomplished the computer shuts down the engine. The computer would also do the regimbaling—the thrust has to pass through the orbiter's center of gravity, and the engine on its gimbals has to be swiveled a little like an outboard motor, because it has to be aimed in a slightly different direction for a one-engine burn than for a two-engine burn. All the astronauts have to do, Ted said, is sit back, enjoy the ride, and remember, when the burn is halfway done, to cross-feed fuel from the other engine's fuel tanks. This is not only to ensure that the double-length burn not terminate from lack of fuel (which might happen if it draws from only one set of tanks) but also to balance the orbiter, so that its center of gravity, and hence the thrust, remains the same.

All this is at once complex and fascinating, as though the relation-ship between the computers and the maneuvering system in some way parallels the connection between the brain and the wing muscles of a bird. But the orbiter, of course, is not an organism but a machine, created by human intelligence. It is a mechanical puzzle the astronauts had to learn to solve. If the training is like a chess game, then the orbiter is the board on which it is played—and its rules are the rules of the game. Hence all the fine points of thrusters, maneuvering systems, and verniers; of inertial-measurement units, auxiliary power units, and fuel cells; of computers, multiplexer-demultiplexers, and restringing, were of supreme interest to the astronauts—even, of course, to Kathy, who would have no direct flight responsibilities. In following the crew through its training, it was impossible not to learn a lot about the orbiter, myself.

At length the burn got under way; the instructors could follow its progress by watching the fuel drain from one set of tanks. Browder looked at his watch, to time when the burn would be halfway over. "They forgot to cross-feed!" he said.

Andy Foster said, "They'll be lucky to have enough fuel in these tanks to make it through the burn."

Ted said he thought they'd make it, but he wished they'd pay more attention to their checklists. He told Andy to put in his notes that they try a single-engine maneuvering-system burn in a week or so, to make sure that Jon learned to balance his tanks. Then he told me that, overall, the crew tended not to use procedures and checklists as well as it should. It was at the basis of McBride's troubles with the maneuvering and thruster systems. "Particularly in an emergency, they try to free-wheel a problem—they don't go to the document and do it step by step," Ted said. "And they leave things out. When that happens, we punish them—say, by leaking a valve they forgot to close." He conferred for a moment with Andy. I had the distinct impression that the crew might not make it to the Honolulu Airport.

Ted seemed to relent a little. "They're all getting better," he said. "Early in training, there's always more anxiety to solve a problem and move on to the next—to cut corners. It's true of every crew I've trained. And we still have plenty of time with this crew. It's better to start slowly and learn the procedures well from the beginning. Crippen did that long ago—he *knows* the procedures, so all he has to do is boldface them." In other words, all Crippen had to do was glance at the boldface headings, and the details under each would come to mind. And the danger, Browder felt, was for a group of rookies to try to put themselves on that level too soon. Mastering the checklists so that they became almost second nature was essential. "It's very easy to miss something," Browder said. One common sort of problem that demonstrated the need for careful attention and thorough familiarity was that one checklist might refer you to another, but the astronaut might then miss a note in the second book to bring him or her back to the first. As a result, a crew might fail to do the latter half of a powerdown. I once asked Kathy if she ever had nightmares about getting lost in the checklists during a mission and not being able to find the right page, and she said, "No, because it happens to me all the time in the simulators."

Browder went on, "So we try to get them to stick with the flight data file—the checklists—and master it." I asked whether the astronauts were given classes in speed reading. "It wouldn't be useful," he said. "There are a lot of abbreviations and lots of implied stuff, so speed reading wouldn't be a help. They must cookbook it the first few times —read every line, do what it says, as if they were following a new recipe—and then, when they have it under their belt, they can scan. Procedures are not like novels; they are not wordy. It's a shorthand type of language, and part of the problem is to teach the language. We expect a lot of homework from the astronauts." Kathy Sullivan once said, "My intentions are good, but sometimes there is more than I can do."

"In another way, a checklist *is* like a novel in that it is a narrative with a deeper meaning," Ted went on. "It's important that they not simply read the checklist to see what steps come next, but beyond that, that they understand the underlying structure of a checklist, its purpose: not only the steps it is trying to accomplish, but why it is logical to do these particular things to achieve this particular end. Reading the checklists, the flight data file, back at home, they can get the gist of it—the logic of what the procedure is doing, and why. 'What does this mean?' 'Why do they want me to shut this bus, or that valve?' That's why, early on in the mission simulators, I want them to take longer over a problem, because I want them to learn it. I want them to follow it. Later, when they know it, I'll throw it back at them—with another problem." That, of course, is the point they were arriving at now.

When the de-orbit burn was over and the maneuvering system engine wouldn't be needed any more, Jon dumped its remaining fuel overboard. Because the tanks were at the rear of the spacecraft, along with three main engines and two maneuvering engines, it was advisable to have the craft's center of gravity farther forward for entry and landing, so that it would fly more like an airplane. Browder watched the monitors closely. In the rear, the thruster system fuel was shut off from the maneuvering system fuel by isolation valves, and if Jon did

not double-check that these were closed when he dumped the maneuvering fuel (something the checklist theoretically reminded him to
do), he could dump the thruster fuel, too. He did it right this time, but
on a couple of earlier occasions he had not followed the checklist and
had jettisoned all the fuel. "The maneuvering system and the thruster
system can really kill you," Ted said. On those occasions, he and the
team had seen to it that the craft completely and irretrievably lost
control of attitude. In punishing erring astronauts, Ted usually liked
to leave a loophole so the crew could find a way out if it was smart
enough—but when all the fuel was dumped, only an irrevocable
punishment fit the crime. Before entry interface (the point four hundred thousand feet above the earth, where the atmosphere is deemed
to begin), the spacecraft has to be tilted up at an angle of attack of forty
degrees, so that the black tiles on its underside take the brunt of the
heat. Without the fuel to do this, the craft would burn up. During entry the orbiter relies on the thrusters for attitude control until it is well
down inside the atmosphere. Andy Foster was spending a lot of time
with Jon in the single-system trainer, helping him with maneuvering-
and thruster-system problems; evidently the extra drill was paying
off, for Jon had never dumped the thruster fuel again.

Randy was making little jabs on the screen with his light pen, and
I asked him what he was doing. Now that entry interface was approaching, he said, Jon was going to start up the pumps that operate
the auxiliary power units for powering the hydraulic systems that control the wing surfaces, the tail flap, the landing gear, and the speed
brakes for maneuvering farther down, and Randy was going to see to
it that one of the three power-unit pumps wasn't going to start. (There
are three parallel and entirely redundant auxiliary power units, each
with its own pump for pressurizing hydraulic fluid in its own pipes
that redundantly control all those moveable elements. Normally they
all work together, but each system by itself can bring the spacecraft to
a safe landing, although the flaps, brakes, and landing gear would
move more slowly.) Jon, who was unaware of this particular gremlin
lying in wait, was inadvertently too quick for Randy; he had already
started the power units before Randy had completed putting the malf
into the computer. Simulator instructors tend to become peevish when
their dirty tricks are circumvented. "He screwed me up!" Randy
bitched. "He ran through the prestart checklist too fast and went right
to 'Start' before I could get my malf in! Jeez! I can fix that!" He jabbed
at the screen with his light pen, and pretty soon an alarm was heard in

the crew station—Randy had failed the power-unit pump while it was up and running.

"We have lost pressure in power unit 3," Dave reported. "There's a message behind that!"

Simulator instructors have a godlike power over the astronauts, and sometimes they wield it with the pique and capriciousness of the Greek deities at Troy. Randy folded his arms and leaned back complacently, a look of Olympian satisfaction on his face. He and Shannon exchanged glances; they seemed to be having a competition to see who could screw up the astronauts the most. They appeared to enjoy each other, and had begun going out together—a romance that evidently thrived on malfs, glitches, anomalies, and nits. The competition that existed between the crew and the team clearly existed *within* the two groups—more so, perhaps, within the team than within the crew, whose interests finally lay in mutual support and interdependence.

Soon the crew was in the radio blackout period, which begins shortly after entry interface; as the spacecraft planes on its gently curved bottom a third of the way around the earth, using the atmosphere to slow itself down, the heat generated by the friction against the black thermal tiles reaches almost three thousand degrees—enough to ionize the air around the craft, preventing radio communications for ten or twelve minutes. In the simulators, during the supposed blackout period, the instructors refrain from making ground calls to the crew, though they listen to their conversation and may even make comments.

During this lull, the team chatted. Ted said that the *Challenger*, atop a Boeing 747, would be flying over the Johnson Space Center that day, on its way from Edwards Air Force Base, where Crippen and the 41-C crew had landed it, to the Kennedy Space Center, where Crippen and the 41-G crew might well be launched in it. Shannon said she wanted to be sure to see it fly over; Randy said he wanted to go, too. "They did that after STS-9," Randy said. "The 747 with the shuttle on top flew slowly over the Center, and everybody was out there, on the big lawn. The 747 dipped its wings."

"And the shuttle fell off, remember?" Ted said. The team would have no time for sentimentality until the launch.

Randy lunged at his TV monitor with his light pen, and soon an alarm sounded in the crew station.

"What's that?" Ted asked. He always liked to have unscripted nits, glitches, or malfs checked with him before they went in.

"It's nothing, nothing at all," said Randy, who evidently was still annoyed about the power-unit ace-out. He may also have wanted to show Shannon who was king of the nits. Dave could plainly be heard saying that they had lost a water loop for cooling, and that he was switching to the redundant water loop. The cooling system for the craft was even more complex than the power units: two parallel and redundant systems piped water through the cabin and the warm electronics, passing on the heat to two parallel and redundant freon cooling systems, such as exist in air conditioners or refrigerators. Each freon system, in turn, passes on the heat to one or the other of two big silvery radiator panels hinged inside the cargo-bay doors, which, when the doors are open, stand by themselves to dispel the heat. If one of these systems breaks down (which Randy had just caused to happen), the other system can provide adequate cooling, although during entry, when a lot of equipment is up and running, some of it has to be powered down. (Because heat doesn't radiate as freely in the atmosphere, there are two other means of dispelling it during entry, but neither is very effective.) Browder said he didn't want to get into doing a powerdown now. Randy retracted his leak. Shannon glanced at him, her eyebrows raised in a smug arch; clearly hers was the defter light pen.

"Just let me know when you are going to do something like that, that's all," Ted said to Randy, and Randy said, "Gee whiz!" Restraining his teammates was not always easy.

"I'm going to foul up their TACANs," Shannon said. Referring to the tactical air navigation system for landing which she had broken before. Ted said to hold off.

A little later, Randy apologized. "I'm sorry about the water loop leak," he said to Browder. "I was just getting itchy. I didn't have enough to do."

Ted said, "That's O.K. You're getting like another member of our team," and he looked at Shannon. Randy seemed to take this as a compliment.

Shannon took the occasion to sneak in her TACAN problem; this time, Dave quickly solved it by referring to the procedures and switching to another TACAN station. He clearly remembered the previous TACAN flub. Shannon looked pleased—pleased with herself, pleased with Dave—and Ted did not chew her out.

Through all this activity on the part of Randy, Shannon, and Ted, Andy Foster had been very quiet, as though he secretly sympathized with the crew. Now his light pen moved to the screen—carefully, precisely, even gently. He jammed the gearbox that operated the speed brakes, the pair of flaps at the rear of the vertical tail (the rudder) which opened to make a wedgelike V to slow the craft as would a drogue parachute. They were now open to their fullest extent, sixty degrees, for the maximum braking as the spacecraft approached Honolulu Airport, just appearing over the horizon. There isn't much the astronauts can do about a jammed speed brake, except hope the spacecraft has enough speed or energy to make it to the landing strip. To make matters worse, there were high crosswinds above Hawaii, further reducing energy. By now the Honolulu Airport was coming well into view on the visuals—Andy told me that it was one of the worst contingency landing sites because it was on an island hardly bigger than the concrete strip, so that if a craft was long or short, left or right, it would land in the water. Later Andy apologized to the astronauts for what he had done; he said he had forgotten that he had also put into the computer very strong winds. Whether his tears were genuine, or those of the crocodile, was uncertain; the maneuvering system that McBride had forgot to cross-feed during the single-engine burn had been Andy's system. Andy seemed to be getting into the swing of things.

Sally suggested that they try to shut the speed brake manually (by switches) instead of remaining in the automatic mode, but that didn't do any good. (Later, in the debriefing, Browder complimented her for the suggestion and reminded the crew that if a speed brake suddenly does unjam and release, they should watch out that they suddenly don't have too much energy and overshoot the landing site.)

"I'm not sure they're going to make it," Andy said, with a deadpan expression.

"Keep your nose down," said Randy; a lower angle of attack, reducing the planing, would conserve energy.

The Honolulu Airport disappeared back over the horizon, which was a bad sign.

The orbiter landed in the water. Even so, Dave, who was playing pilot to McBride's commander, had lowered the landing gear—not recommended when ditching in the ocean. "Shoot," said Jon when Dave had confessed. "I thought we might have made like a big boat and coasted right on in."

"They thought you'd want to go for a swim after your arrival," Sally said.

"I'm sorry we took that much energy away from you, with the crosswinds and the speed brakes," Ted said. "I'd have liked you to have gotten that one landing in, in case you ever have to land in Honolulu." I wasn't sure how sorry Ted was, either.

"Were you looking for an aircraft carrier out there, Jon?" Andy asked.

"Yeah, but they were all out at sea," Jon said. "But we landed nice and soft."

"There were some surfers out there. They'd have opened the hatch," Sally said.

I wondered how much of the astronauts' seeming good spirits masked anxiety, an edgy feeling at the back of the mind that something might go wrong on the flight. In a way, the sims were parodies of what could happen, and therefore the source of a good deal of nervous humor. The humor, though sometimes black, was basically good-natured and upbeat. Possibly it involved a process of building confidence in themselves and the spacecraft, as well as in the people who designed and serviced it. The belief in success in a dangerous enterprise is often accompanied by a sort of forced good spirits. Equally likely, though, the astronauts were all highly motivated people who were doing exactly what they wanted to do most, and there is always joy in that.

A day or so later, Dave Leestma and Kathy Sullivan practiced their EVA in the weightless environment training facility, the WET-F. They had done this about three times before, and they would do it several more times before launch. Under water, an astronaut can be neutrally buoyant, neither falling nor rising, simulating the conditions in space. The WET-F is an enormous pool, about seventy-five feet long, fifty feet wide, and twenty-five feet deep—big enough to contain on its bottom a replica of the orbiter's payload bay and other equipment. It is inside a large round building with the dimensions of an old roundhouse for locomotives, and which once held the centrifuge, a metal boom fifty feet long with a gondola at one end, in which astronauts were whirled to simulate the gravity forces exerted on astronauts during launch and entry. The centrifuge was abandoned after the Apollo program because astronauts returning from orbit in the shuttle rarely pull more than 1.5 G's. There are many other ways in which astronaut training has changed since the Apollo days, usually in the direction of being less dramatic. Mercury, Gemini, and Apollo astronauts used to go through survival training in various parts of the world, in case of a forced landing, in, say, a desert or a rain forest; accordingly, those astronauts went to exotic spots like jungles in Guatemala or deserts in Mexico and learned to do such things as catch and roast iguanas. Because the shuttle is much more likely to make it to one or another of the contingency airfields, replete with Coke machines and cafeterias (or else ditch in an ocean), this sort of survival training has been abandoned.

The pool in which Dave and Kathy would simulate the EVA makes a big greenish-blue rectangle on one side of the floor, like a giant carpet. (NASA, a stickler for safety on earth as well as in space, overchlorinates.) The big round room is warm and humid; the pool is heated to about ninety-six degrees, just below body temperature, because the scuba divers in attendance on the astronauts are under water for two or three hours at a time, and they don't have the luxury of dry, warm space suits, as the astronauts do. The other side of the room from the pool was filled with lightweight mock-ups of all the different equipment astronauts in space had used, or would be using, in the cargo bay—a record of past and future flights. There were mock-ups of the satellites and payload assistance module rockets that astronauts on previous missions had deployed; there was the big platform that held the Solar Max on 41-C; and there were pallets the 51-A astronauts (scheduled for the flight after 41-G) would use to retrieve two errant satellites. All these mock-ups were made of wire mesh, so that they could easily be lowered to, and raised from, the bottom of the pool. The mock-up of the orbital refueling system, which Dave and Kathy would be using, was already down at the bottom, inside the mock-up of the cargo bay; it was about the size and shape of a barrel with an opening below a hood. Dave and Kathy would have to learn to reach in and, in an enclosed space, do the intricate job of opening a couple of valves on a fuel pipe that had never been designed to be opened by astronauts in space, and then rig a connection to a fuel tank. The pipe, one of a cluster of four, was identical to a similar fuel pipe on Landsat, a satellite that has been successfully photographing the earth for several years; its pictures, taken in a number of wavelengths, are valuable to geologists, agriculturalists, forest experts, and a wide variety of scientists and engineers. If astronauts could refuel the satellite with hydrazine, Landsat would have many more productive years. The trouble was that Landsat, and most other satellites, had not been designed to be

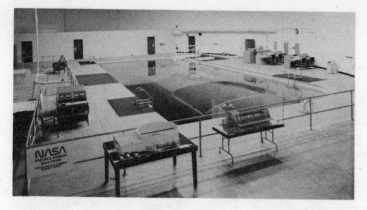

The weightless environment training facility, or WET-F, in which astronauts practice extravehicular activities, or EVA's. The pool is about seventy-five feet long, fifty feet wide, and twenty-five feet deep. The shape of the orbiter's cargo bay can be made out under the limpid, tropical, highly chlorinated water of the pool.

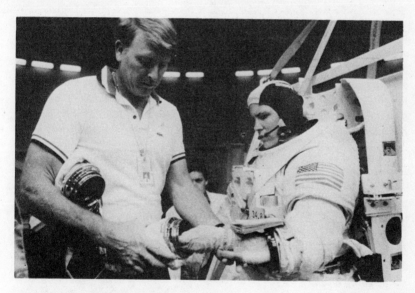

Jon McBride helps Kathy Sullivan put on her spacesuit gloves just before she is hoisted into the WET-F. Jon, who would be in charge of overseeing the EVA from inside the orbiter, is learning some of the ways he would have to help Kathy and her partner, Dave Leestma, prepare for their spacewalk during the mission.

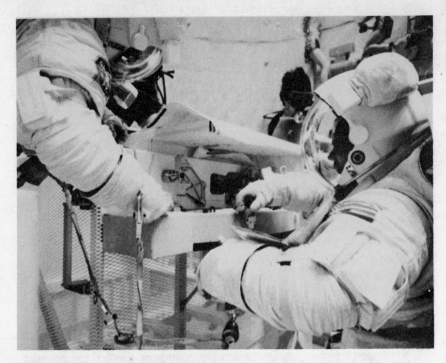

Kathy Sullivan (left) and Dave Leestma practice with a mock-up of the orbital refueling system, under about twenty-five feet of water, where— because of a balance between the air in their suits and weights attached outside their suits—they are neutrally buoyant, in an attempt to simulate weightlessness. The refueling-system mock-up is largely made of mesh so that it can easily be submerged. Photograph taken for NASA by Otis Imboden.

Robert Crippen, the 41-G commander, gets ready to put on mask and oxygen tank to observe Kathy and Dave practice their EVA. Even though he and Jon McBride, the pilot, would not be going on the EVA, they had to be thoroughly familiar with every facet of it.

serviced by astronauts, so that Dave, assisted by Kathy, would have to use special procedures and special tools to make a water-tight, or hydrazine-tight, connection between a supply tank and an empty Landsat fuel tank—or rather, between tanks like them. Once it was demonstrated that satellites could be refueled in space, a whole new area of service for the shuttle would open up.

Dave was sitting by the side of the pool in his water-cooled undergarment, a sort of union suit laced with little tubes like capillaries that circulated water around his body and then to his backpack, where the heat would be dispelled and the water recirculated. Various tubes and hoses protruded from him; there is probably no rattier sight in the solar system than an astronaut in his water-cooled undergarment. Dave was checking the tools he would be using, about eight of them, which were lined up on the edge of the pool; they had all been made at the Johnson Space Center, and they included a tool that resembled the scissors a surgeon would use for cutting in an enclosed space (a piston that pushed a blade against an anvil), a flashlight with a flexible fiber-optic neck with a dental mirror at the end for illuminating tight places, and six other unique tools designed for this operation. Tom Grubbs, a NASA engineer who had worked on the tools, looked on anxiously as Dave assessed them. Dave complained that they weren't the same length; he said they should be. On the STS-11 mission, Lieut. Col. Robert L. Stewart, who had used similar tools on an earlier version of the orbital refueling system, had complained that they were too long. Since then, some of them had been cut down, but not all of them. Even though Tom Grubbs assured him that the flight articles would be perfect, Dave—perhaps the most critical member of the crew—did not look overly confident.

He wriggled into the upper part of his spacesuit. Called the hard upper torso, it was solid plastic with the life-support system bolted to it, and at the moment it was attached to the upright part of a platform on which Dave would be hoisted by a crane into the water. (Similarly, in the orbiter, his and Kathy's hard torsos would be secured to the wall inside the airlock, and they would wriggle into them in the same way.) Once inside, with his arms and legs sticking out, he looked like a turtle in a shell— a shell, moreover, three sizes too big, so that Dave (or anyone else in his position) looked scrawny. Astronauts do not look their best while putting on their spacesuits. Technicians helped him pull up a pair of spacesuit pants that had a metal ring in place of a belt; it locked to another metal ring at the bottom of the hard torso. Gloves

locked to wrists in similar fashion, and so finally did the helmet. All
NASA's pants, gloves, and torsos were different sizes, and the astronauts dressed in the ones that fitted best; once Dave, who is one of the
shortest of the male astronauts (he and Kathy wear the same sizes),
picked a pair of pants that were too big, so that he fell down inside his
suit and couldn't see out.

While Dave was getting ready, several divers with oxygen tanks,
goggles, and flippers, who would assist him, got into the pool. Though
they had probably done this hundreds of times before, their spirits
were as high as anyone's going for a swim; they splashed and blew water at each other. An ambulance arrived; there is always one parked by
the pool when suited astronauts are in it. Though there has never been
a serious accident, there have been some close calls, and two divers are
assigned to attend each astronaut at all times; a third diver for each
astronaut serves as a sort of gopher. The astronauts in their suits cannot swim; the divers have to move them, fetch for them, and be on the
lookout for trouble. While training for Crippen's 41-C mission, James
D. Van Hoften had a hose for oxygen pull out of his suit; the divers had
him up a ladder, out of the water, and his helmet off in thirty seconds.
And on another occasion Crippen's spacesuit boot ruptured; he too was
hauled out. Leestma told me once that he didn't have claustrophobia in
his suit under water because of his confidence in the divers; in an
emergency, they could yank off his helmet and stick a scuba tube in his
mouth in almost no time.

I joined Jim McBride, a tall, quiet engineer who was in charge of
EVA training for the 41-G astronauts, in a control area, where there
were several rows of chairs facing two TV monitors; in the front row, at
a desk, sat a technician with a microphone who would act as the capsule communicator. In addition to the divers attending the astronaut,
two more operated underwater TV cameras—a total of five divers.
(When there are two astronauts in the pool, as is the usual case, there
can be as many as ten divers.) The view from twenty-five feet down
was bright and suffused with the blue-green of chlorine; the astronaut, his equipment, and the cargo bay were white; the divers, in dark
blue bathing suits, were well tanned. They stuffed heavy lead weights
into pockets attached to Dave's legs and torso—fifty-five pounds of
them would be needed to counteract the atmosphere in his suit; otherwise he would pop to the surface like a cork and shoot into the air. They
wanted to make him neutrally buoyant, not only so that he would stay
in the same place, but also, of course, to simulate weightlessness. It

took some care: they had to do it in such a way that the weights were balanced with respect to his body, so that one end didn't go up and the other end down; accordingly, they laid him flat in the water and added weights here and there until he balanced like a pair of scales. It was a virtually impossible task, because the air inside his suit kept shifting around. Indeed, the WET-F has its limitations as an analogue of weightlessness. It is very much a part of this planet, and therefore— however neutrally buoyant an astronaut is rendered—objects he or she uses that are heavier than water, such as a metal tool, will drop to the bottom of the pool, in an unspacelike manner. And even a neutrally buoyant object, if given a shove horizontally, will stop very soon because of the density of the water; in space, of course, it would go on forever. Another way of simulating weightlessness is to fly great parabolic curves in a KC-135 airplane, which in many respects provides an experience much closer to being in space; however, each period of weightlessness, which occurs as the plane flies over the top of its parabola like a rollercoaster swooping over a hill, lasts less than thirty seconds—hardly enough time to practice most tasks, let alone an entire EVA. Its chief effect is to make astronauts and instructors nauseated. Weightlessness is so different from anything known on earth that, as was the case with the manipulator arm, it has to be simulated in more than one way, each getting at a different aspect, just as a surveyor measures the distance to some inaccessible point by sightings from a variety of positions.

Although tethering was not necessary in the WET-F, still everything and every astronaut in the pool was tethered, simply to accustom the astronauts to the procedure. Dave himself was tethered by the waist to a cable that ran down one side of the payload bay, along the hinges of the huge doors (there was a similar cable, for another astronaut, running down the other side); as he and Kathy would in space, Dave could move from one end of the payload bay to the other by pulling himself along handholds just beneath the long hinge, as if he were on a dog run. A reel took up any slack in his own tethers. Smaller tethers and smaller reels secured all the tools; some of the smallest ones fitted inside folding pouches called caddies that had tethers and reels built into them. Tools and caddies could be mounted on a metal bracket called a miniwork station that fitted on the front of the astronaut's suit. Astronauts were forever getting tangled up or reeled in to places they didn't want to go, like fish. The only astronauts who are ever untethered on an EVA are those who ride around on the jet-propelled manned maneuvering units. After George Nelson's return from the 41-C mission, on which he had wrestled the Solar Max with his maneuvering unit, he said to Kathy Sullivan, a little condescendingly, "You have only one EVA, and you're tethered." (There is a friendly rivalry between crews, just as there is between crews and their instructor teams.) Kathy, who had just arrived in the pool, wearing a black bathing suit instead of a spacesuit, repeated this remark to Dave after his WET-F sim was over. Dave replied stolidly, "We'll take the tethers." He had no intention of sharing the fate of the astronaut in the film *2001: A Space Odyssey* who drifted away from his spaceship— something that almost happened to a Soviet cosmonaut several years ago.

Dave slid down the wire to the rear of the cargo bay and stood in front of the mock-up for the orbital refueling system. As he would in space, he stood in a pair of foot restraints like the toe grips of ski bindings, which would hold him firmly in place while he put pressure on the tools. The restraints had to be adjusted because he was so short— indeed once, when he wasn't using them, the divers had to bring him down a stool to stand on. Dave did not like the foot restraints; sometimes they broke, and sometimes his feet just came out so that he found himself floating slowly upward. Kathy, who was not held down with foot restraints, hovered alongside Dave, trying to peer over his shoulder and occasionally pretending to snap a photograph of the progress of his work, as she would in space.

As Dave set to work on the refueling system with the tools, several technicians moved closer to the TV screen. "If he likes them, NASA will buy them," Jim McBride told me. Tom Grubbs sat expectantly on McBride's other side. Dave's first task was to cut a couple of safety wires, one of which secured a cap on top of the fuel pipe opening, and another of which secured a nut just inside. The wires were almost impossible to see—the fuel pipe, and the other pipe endings, were grouped together inside a metal container; although it was open at the top, Dave couldn't get his head near it because the whole apparatus was enclosed inside a metal hood. He had to work largely by feel. The goose-necked fiber-optic flashlight with the mirror at the end allowed him to see the wires—Dave had invented the flashlight himself, and Stewart had used it on the last mission and found it valuable. Astronauts can have an enormous influence on the design of the tools they use in space, and Dave, who had a constantly questioning nature, was particularly concerned with design—sometimes the tool makers wished he was a little less interested. Still, Dave persisted; if anything didn't work in space, he reasoned, he would be the one to suffer. Kathy held the flashlight for him while he attacked the wires with the clippers. The wires were very tight; Dave had to pull them loose with a notch at the end of the tool, so he could get the clippers around them. The notch had been Kathy's idea: "It's Kathy's crochet hook," Tom Grubbs told me. Whatever male chauvinism persisted at the Space Center seemed to be gentle and antiquated.

With the first wire loose, Dave put the cutting part of the clippers around it and pulled a trigger by the handle; the trigger caused the sharp blade to move forward, severing the wire. The block at the end had broken off when he squeezed the trigger. "Rats," Dave said, simulating the correct way for an astronaut to express annoyance.

"Damn!" Tom Grubbs said. "I hate it when equipment breaks— too many people want to know why."

"That's what these tests are for," another technician said, trying to buck him up.

A technician sitting behind Grubbs said something about no matter how hard they tried, they could never make the tools astronaut-proof. Dave, it turned out, was totally innocent; when the broken tool was brought to Grubbs, he declared that the metal at the end was too brittle, and that the rod that pushed the blade against the block was a quarter of an inch too long, so that too much force could easily be exerted on the block. Dave told me later that the clippers had been cut down, which he thought was unnecessary, and that the block had broken off where it had been welded back on. "There's an old NASA philosophy—if it works, don't mess with it," Dave said.

Dave went on with his task, delivering a commentary on what he was doing the whole time. "Crewmen like Dave really get involved with design," said Tom, who had recovered his composure. "He's making comments now on things he wants to help me with." Dave unscrewed the cap on the end of the refueling system's fuel pipe with a long tool with a clamp on the end that he stuck down inside the box the pipes were in; this tool, he said, worked fine. Next came a tool called the ball valve assembly, a pipe wide enough to fit around the fuel pipe, screwing on to it where the cap had been. Dave said he was pleased with the modifications that had been made since he had last used it— it was shorter, making it easier to maneuver in the small space under the hood, and it could be screwed on with fewer turns. Dave said he was glad he was held down with foot restraints, because if he wasn't, all the turning of screws would cause him to rotate away into space; he was glad, he said, that fewer turns were needed to lock the ball valve into place, because twisting in spacesuit gloves is hard on the fingers. Nelson, who had done a delicate repair job on the Solar Max, found afterwards that all his fingernails were bent halfway back; and Robert Stewart, after doing the refueling system work on STS-11, had found his fingers were bruised. To avoid what was known about the corps alternately as "Nelson's finger" or "Stewart's finger," Dave had selected a somewhat larger pair of gloves, and he kept his nails closely clipped; he never had any trouble. When the assembly was in place, he turned a golden lever on one side of it that opened the ball valve. This was simply a ball in the middle of the pipe that could be rotated by the handle: in one position, the ball blocked the pipe completely; in another position, a hole through the ball lined up with the pipe so that it was open. He complained that the handle's operation was too stiff. With the valve open, he inserted another, narrower tool through it and down into the fuel pipe to remove a nut that held a seal in place. He shut the ball valve, for with the seal out, there was only one other seal deeper down in the pipe—because hydrazine is highly toxic, it was deemed advisable at all times to have two seals between the fuel and the astronauts. (In the WET-F, of course, the experiment was not loaded with hydrazine, but it would be in space.) He inserted the last tool—the multipurpose tool—which opened the seal farther down while keeping two seals within itself closed; this tool, which would be left permanently in place, had a hose at its upper end that Dave connected to another pipe. When that was done, the connection was complete. The entire job had taken about an hour. There was a considerable jumble of tools and other debris around him. "It's hard to keep track of all those tools," Dave said. "Kathy is here; we'll give her the clean-up detail."

"Don't make Kathy mad," the technician acting as capsule communicator said.

"There are a lot of bad vibes coming from Kathy now," Dave said.

"Careful," said the capsule communicator. "We don't want her to pop her helmet." Fortunately, she wasn't wearing one.

Hairy Things

During the week of June 11, about seven and a half months into training, a number of loose ends were finally nailed down. No longer need
the crew address the team as "Poughkeepsie" or "Moundsville"; or the
team address the crew as "Potemkin." They would use the *Challenger,*
allowing the *Columbia* to remain at the Rockwell factory in Palmdale,
California, for a complete refitting—it would even get a heads-up display. The launch date was set for October 1, about three and a half
months later, and the *Challenger* would land, after an eight-day mission, at the Kennedy Space Center. Mission Control would stay in
Houston.

And the loosest of loose ends was in place, for Crippen had been
training with the 41-G crew for about two weeks. All hands seemed
glad to have him. "It's like adding the bonding agent to the mission,"
Browder told me, referring to a strong glue. They were, in fact, a crew
now, where they hadn't really been one before—or, rather, with Crippen the crew had taken on its final form. "Now that Crippen's back, we
feel a lot better," Jon McBride said. "He fills in the gaps. He keeps us
from heading in the wrong direction." Kathy Sullivan said, "He's a fo-
cal point we didn't have before. Earlier, we relied on each other and
tried not to bother Crip when he was working on the 41-C mission.
Now it's diametrically opposite—we tell him everything; we make
sure he misses nothing we've been doing."

Crippen slipped in so easily that one of the crew said, "We sized
the hole for him to fit in perfectly." The integrated sim that had so worried Browder—the one in which the crew stood in for the 41-D crew—
had come off, Ted told me, "swimmingly." Beforehand, Crippen, too,
had been a little worried. "I wondered a little about doing that sim
myself," he told me later. "I don't want to kill people, but there's
nothing like a little practice under fire, if only just to demonstrate that

they *could* do it." Fortunately, he was able first to get in three stand-alone ascent sims with the crew; with those successfully behind him, he told Ted, just before the integrated sim, "We could do this standing on our heads." In the course of the integrated sim, the crew only made one error, and it caught the flight controllers in an error. Crippen signed up the crew for a couple of more integrateds when the 41-D crew, which now was "next up," as the crew about to go into space was known, would be away. Ted had told me before that he expected the training in some respects to be four to six weeks behind when Crippen returned; now he told me that they were no more than two weeks behind, as judged by lessons, such as contingency aborts, which they would be doing this week, but which ordinarily would have been done earlier. "We need a commander to do them," Ted said. The minute I entered the simulator control room, I could sense a change. There was less chatter. Because Crip, in the commander's seat, did most of the flying, he also did most of the talking. Sally Ride told me that having Crip in the commander's seat made a big difference in the way business was carried out in the crew station, for it meant that the other crew members—all of whom at one time or another had stood in as commander—went back to their own jobs and stayed there. "The business of people turning to me to find out how Crip wanted things done happened until the day Crip came back," Sally told me. "But now we don't need a Crip stand-in any more. It took only a day and a half after Crip came back before it felt as if he'd been here all along. It's great to have a commander! The big difference is that there is an obvious leader." And Kathy said, "Things go better now, because he has the experience to get things done efficiently. He is the fulcrum during ascent and entry; he does the flying, and the rest of us help out. He delegates the engines to Jon and the checklists to Sally. I go to the back seat, where I have no particular role, and Dave goes below deck."

And Crippen was glad to be with the crew. He was pleased with its progress in training; after their first few sims together, he told Ted he was surprised at how well everyone knew what they were supposed to do. "He said he wondered what they needed a commander for—though of course there is no way they could fly without him," Ted told me. And Crippen said to me, "They're in much better shape than I anticipated. I felt almost as though I'd been with them the whole time; they did things just the way I wanted them to." He had to tweak things a little for his personal preferences, such as when he wanted the wheels lowered for landing and the use of only one of the mission specialists—

Sally—to help with checklists during ascent and entry. (Some commanders prefer using two. "I think one extra pair of eyes is enough," he told me.) He even brought back some new ideas from the 41-C mission: he found that hand-held microphones worked better than earphones, so the crew immediately set about using the hand-held variety and were glad of the switch. All in all, Ted was what he called "thumbs up" on the experiment of the commander joining the crew late, and so was Crippen.

There were a couple of problems that followed from Crippen's return, though. One was that at first the four other crew members had a natural tendency to sit back during sims and see what he wanted them to do, instead of taking the initiative. It was an after-you-Alphonse situation, according to Ted, for while everyone else held back to see how Crippen wanted to do things, Crippen in fact wanted to see how the crew was doing them. Sally, who had flown with Crippen before, came out of this period of reticence quickly, by the end of the first day. Kathy and Dave shed this natural hesitancy perhaps a day or so later and were charging around, doing what needed to be done as they had before. Jon McBride, however, was taking a little longer to be completely at ease, which was, in Ted's opinion, absolutely right, because as pilot his job was more intricate than the others', and was more interdependent with Crippen's. "Jon is quite meticulous—he wants to make sure things are to his understanding before he commits himself," Browder said. Of course, if Crippen had been there all along, he and McBride would long since have been functioning as a unit. Crippen had told me earlier that developing a mutual trust was about the most important thing the simulators did, and it could be that the effect of the commander's absence on the pilot's integration into what had to be a close two-man partnership was the one area that bore watching. Ted was confident that Jon would get over this entirely understandable reticence (he was, after all, teamed up with the most experienced of commanders) very soon. And Jon was getting more and more proficient all the time, Ted told me. Although he still had a little way to go with the maneuvering and thruster systems, he had made some giant strides, and Andy Foster was having to reach out a little harder all the time to fool Jon in these areas. Ted and Andy were looking forward to working some more in the simulator with Jon, but they would have to get Crippen out of the way first.

Indeed, one problem that stemmed from Crippen's return was the same problem that initially had made Browder glad to have the other four astronauts by themselves for so long: Crippen was such a skilled orbiter pilot that without meaning to he sometimes thwarted the training of his own crew by jumping in and solving problems himself. And Crippen had been the very first to recover from the period of shyness after his return. Now, if Ted or Andy tossed Jon a maneuvering system problem, Crippen might well grab it and solve it before Jon knew there was anything wrong. "With Crippen there, we have a harder time fooling the crew," Browder said. "Crippen has seen about every training scenario there is." Furthermore, Crippen has almost total recall for any scenario or anything else he has ever heard, read, or experienced, and can draw on this reservoir instantly. He is quick to notice things. He can rapidly scan the spacecraft telemetry data on the computer screens and in the columns of numbers spot a problem inserted by the instructors in its earliest phases, long before it has set off an alarm. He can tell a sensor problem from a real problem a mile away. However, Browder has observed over the years that as Crippen gets more proficient, he is becoming more aware of the need to restrain himself so that others on the crew can be trained; Ted felt he was much more aggressive in the simulators during training for his first two flights than his last two. Nowadays, having Crippen aboard was like having a senior instructor in the cockpit. However, it was not Crippen's nature to stand back passively, as a senior instructor might, and sometimes the temptation to rush in and solve a problem that was stumping someone else still proved irresistible. Even if he did restrain himself, he couldn't restrain his facial expressions—and Browder once told me, "Crippen gets more mileage out of a raised eyebrow than anyone else." In the cockpit, all a crew member who was working on a problem, and who wasn't sure the solution was right, had to do was to look at Crippen and—no matter how virtuously silent the commander was—he or she could read the answer in the purse of Crippen's lips or in the distance his eyebrows were from his eyes. Browder and his team had to plan around Crippen. "To prevent him from giving in to his impulse to solve a problem or give a hint, I'll divert Crip," Ted said. "I'll tell Shannon to get Crip out of our hair with a malf in the computers. We're fairly successful at that. It's fair game, I think. But Crippen does sometimes manage to do two things at once. He understands what we're doing, though, and he plays along; even if he finishes the computer problem, he'll leave Jon alone."

On Tuesday, June 12, the crew was scheduled to practice ascents, aborts, and contingency aborts. An ordinary abort is an emergency landing at an airfield, possibly across the Atlantic, or at any of the dozen fields around the world where NASA has arrangements. Safest, of course, would be the dry lake bed at Edwards Air Force Base in California. A contingency abort is a forced landing not at an airfield but anywhere else, most likely an ocean. Because ascents are the trickiest and most dangerous part of a space flight—more so even than entry— aborts are more likely to occur then; for that reason more time is spent on ascents and aborts in simulations, where the two go together like coffee and cream, than on any other phase of the mission. Perhaps half of all time in the mission simulators is devoted to them; if Crippen or any other commander is faced with a choice of what to do in sim, they will almost invariably pick ascents and what can go wrong with them.

In a normal ascent (rarely practiced to completion in the mission simulators), the three main engines in the tail, which draw their fuel from the huge external tank attached to the orbiter's underside, ignite first. The spacecraft, still attached to the ground, vibrates and lurches forward, for the main engines are somewhat higher than the entire shuttle's center of gravity; the vibrations and tipping are faithfully duplicated in the crew station. Combined, the main engines provide about 1.2 million pounds of thrust at launch. Three point eight seconds later, the two solid-rocket boosters, attached to either side of the external tank, ignite, administering, according to some astronauts, what feels like the kick of a horse (not faithfully duplicated in the crew station). The two provide six million pounds of thrust, and thus give the main push off the ground. Instantly the spacecraft begins to climb into the air. Soon after clearing the tower, while the thrust is its greatest, the orbiter, rising straight up faster and faster, rolls around its vertical axis so that the top of the fuselage is aimed toward the craft's eventual trajectory, which is to say in the same direction as its angle of inclination to the equator. Accordingly, when the spacecraft performs its next maneuver, a slow pitching over of its nose toward its future orbit, the craft will be flying upside down. Approximately two minutes into its flight, the two boosters drop off. (They float down to the ocean on parachutes and are retrieved by tugs for reuse.) The craft is now twenty-eight miles high, twenty miles downrange, and flying at a little over three thousand miles an hour. The main engines continue to drive the shuttle to an altitude of sixty-eight miles and a velocity of about 17,440 miles per hour, close to orbital speed. Eight and a half minutes after launch, the main engines cease—an event called main-engine cut-off. When a spacecraft reaches that point, astronauts, ground controllers, instructors, and everyone else breathe more easily, for it is in the period up until main-engine cut-off, with all those million pounds of thrust shooting from state-of-the-art engines, that aborts are most likely to happen. After that, the external tank, with the orbiter beneath it, is jettisoned (it will lob across Europe or Africa and break up in the atmosphere above the Indian Ocean). The small engines on either side of the tail—the orbital maneuvering system engines—give the spacecraft its final boost into orbit, which it reaches some forty-five minutes after launch, halfway around the earth, still flying upside down. Accordingly, when the meaneuvering system engines fire again to circularize the orbit, and when the payload-bay doors are opened, the instruments inside will be looking down at the earth.

This, basically, is the normal flight plan, though in the simulators the astronauts fly a variety of departures from it to practice aborts. Aborts are usually caused by a failure of one or more of the three main engines; they vary in their methods and landing sites depending on how many main engines are lost and at what point in the trajectory they fail. The main engines, using liquid oxygen and liquid hydrogen, are more complex than the solid-rocket boosters and therefore, at least at the time of 41-G's flight in 1984, were considered more likely to fail. Conversely, the boosters are roughly as simple as a keg of gunpowder, and then, if not now, were considered less likely to fail, once ignited. Again, as with a keg of gunpowder, *if* they failed after ignition, nothing could be done about it. Hence, none of the aborts begin until after the separation of the boosters at two minutes. If any or even all three main engines fail before that point, the boosters most likely have enough power to carry out the first of the aborts: they fling the spacecraft high enough into the air, and fast enough, so that it can turn around and return to its launch site (hence, it is called an RTLS)—or rather to the landing strip not far from the launch pad at Kennedy. (However, if all three main engines fail at lift-off or shortly thereafter, the spacecraft cannot do an RTLS, but the big boosters would at least hurl the orbiter far enough to land in the water.) Assuming that one or two of the main engines are still running after the separation of the boosters, the shuttle will head on out over the Atlantic to use up fuel until only enough remains to turn around and fly back to

Kennedy; the turn is made by the orbiter—flying upside down—pitching its nose slowly downward, its big wings slowing the craft and plowing it around, until the craft, having made a loop, is flying right side up, in the direction of Kennedy. (One of the reasons the spacecraft heels over and flies on its back is to make the RTLS maneuver possible.) After reversing direction, the main engines cut off and the external tank separates; then the orbiter glides on in to Kennedy as it might to an ordinary landing.

There comes a point farther along on the normal trajectory, though, where if the orbiter loses one or more main engines, it can no longer make it back to Kennedy and will instead have to fly on across the Atlantic to an airport in Africa or Europe. Depending on how highly inclined the trajectory is to the equator, the transatlantic aborts might land at air fields in Dakar, Spain (very likely Zaragoza), or Germany (very likely Frankfurt). When the opportunity to do transatlantic aborts has passed, the spacecraft can make it to orbit; if the emergency presents a continuing problem—say, instead of a failed main engine there is a cabin leak—the spacecraft would do what is called an abort once around, in which it would orbit the earth, returning to its launch site or to Edwards. Or it might do what is called an abort to orbit, in which it would go to orbit and stay there while the astronauts and Mission Control evaluated the problem. On the way up, the astronauts pass various milestones that relate to abort capabilities, and the capsule communicator is apt to tick them off. After the RTLS capabilities end, there are two-engine and one-engine transatlantic abort capabilities, and two-engine and one-engine abort once-around or to-orbit capabilities; these overlap so that in theory the astronauts almost always have an emergency route to safety. Though the overlapping nature of these aborts looks good on paper and is theoretically what NASA calls a "confidence-building factor," NASA recognizes that situations might arise (a loss of two or even all three main engines, say, at a point where the spacecraft couldn't make it back to Kennedy or to any airport overseas) when the orbiter would have to ditch in the ocean. These are called contingency aborts, and they are the riskiest of all. Browder told me he planned to do some contingency aborts, as well as ordinary aborts, that day.

The astronauts were already in the crew stations; they were using the motion-based simulator, which tilted the cockpit on end for launch and then gradually brought it to the horizontal as the trajectory leveled off; the visuals out the windows were of the launch tower. They felt as if they were sitting in reclining chairs. "I wonder if we could get our offices to do this," Sally speculated.

Ted Browder introduced me to an instructor I hadn't met before, Steve Williams, a contingency-abort specialist. (Sometimes these roving specialists joined the team for specific lessons, as Jim Arbet, the expert on the manipulator arm, had done.) "First run, let's try two engines out at six thousand feet per second," he said. To flight experts, a velocity can be instantly correlated with a point on the ascent trajectory—in this case, the spacecraft would be at an altitude of two hundred twenty thousand feet, ten miles downrange from the Cape. During the ascent, the launch tower slid backwards in the visuals, and as the five big rockets—two boosters and three main engines—pulsed and pounded, the motion-based simulator jounced up and down. So did Browder, Shannon, Randy, Andy, and Steve Williams, sitting in their chairs at their consoles in the control room, which of course rested stably on its foundations. As always, the team was empathizing fully with the crew. As the shuttle rose, about the only sound from the spacecraft was the sound of Crippen humming in a nonchalant fashion. Williams, acting as capsule communicator, informed him that he was at the point where, if he lost one engine, he could make it across the Atlantic—what he called "two-engine transatlantic capability." That, as it happened, should have been a good place to be, because Crippen reported that one engine had failed on him. His instruments, however, were wrong. Williams said that, according to the ground's monitors, he had lost *two* engines, which meant he had to try to do a return-to-launch-site abort. As an RTLS with two engines out is virtually impossible, Williams said, they would most likely have to ditch in the sea. By attempting an RTLS, though, the crew would at least ditch closer to the Cape and its rescue forces. Sally Ride, in the middle seat, rummaged for the right checklist. "The RTLS checklist means you're going to live," she said bravely when she found it. She found the place that most nearly fitted their predicament and read along, following a pathway marked in blue. (In emergency checklists, routes to follow for different contingencies are marked in different colors.) Soon Crippen pitched the orbiter's nose down so that it could do its long, braking tumble, at the end of which it reversed its direction and was headed back toward the Cape. Almost immediately, though, something went wrong: the spacecraft lost control, and Browder ended the sim. "I'm not sure I would have survived that one," Crippen said, and Browder said they'd try to repeat it. This was what Browder called a "sim crash," one of those words like "kill" or "die" that astronauts and instructors toss around indiscriminately—and nervously—in the insulated safety of the simulators. A sim crash means that something has gone wrong with the simulator's computer hardware or software; the spacecraft goes out of control when there is no reason that it should have. Browder told me that the term evolved from something that made the spacecraft crash to the idea that the simulator crashed. "It conjures pictures of little pieces lying all over the concrete," he said. "It means The End. To me, it always means that we're one lesson behind."

The astronauts and the instructors quickly reset their switches for another launch. Soon after lift-off, one engine shut down, and then another, just like the last time, only this time Williams shut down a different pair, in the hopes that perhaps the simulator wouldn't crash. But the astronauts lost control again, for unknown reasons; as Crippen pitched the spacecraft down for the RTLS, it apparently kept on pitching long after it should have leveled out and continued home to the Cape. Rather than ride it out, Crippen asked Browder to freeze the sim and do a data dump, by which he meant to record all the data at the point the trouble had started; clearly the computer experts would have to do a repair job, and they would need the information.

As they reset for the next run, Sally said, "One way you can tell you are in a simulator and not in the orbiter is that all the abort switches are worn, and all the nominal ones are new."

It was proverbial among astronauts that a space flight was a snap compared to a session in the simulators. Some old NASA hands were a little worried by this attitude. Eugene Kranz, who had the reputation for being an old firehorse because he had been the chief of the Flight Control Division during the Apollo program and the lead flight director for many of its missions, told me that, by and large, the shuttle astronauts and flight controllers had not been battle-tested the way their counterparts in the Apollo days had been. Prior to 41-G, there had been difficult moments with the shuttle, such as the loss of a fuel cell on STS-2, and a computer failure on STS-9, but in Kranz's opinion these episodes were not life-threatening. The old Apollo astronauts, on the other hand, had had some genuinely close calls, among them a computer failure during the descent to the moon's surface on *Apollo 11;* a lightning strike of the spacecraft just after launch of *Apollo 12,* which caused the command module to lose its guidance platform and part of its electrical system; and, of course, *Apollo 13,* when an oxygen tank ruptured on the way to the moon, which resulted in a cancellation of the lunar landing and a return to earth around the moon, which was touch and go all the way. "Today, we have a generation of people who have not been severely tested," Kranz told me, "and such testing is the final maturing in the learning process. We have fine people, but until they are severely tested, they won't know how good their knowledge and skills and toughness have to be." Kranz did not seem to think that shuttle astronauts would necessarily always lead such a charmed life. "High-tech accidents, such as the one at Three Mile Island, or *Apollo 13,* are unpredictable and unavoidable," he went on. "We have to do everything possible to avoid them—but when they occur, we have to make sure we can do our best. The training program gets us ready, but until it happens, we don't know if we are ready to the core." .

Next time, Williams failed the third and final combination of two engines. For a moment Crippen seemed to be losing control—the spacecraft was yawing dangerously—but he fought it successfully. "We'll ditch in the Bermuda Triangle," Crippen said, referring to a part of the ocean where mysterious things are apt to happen. Browder told me that although NASA practices in a fairly upbeat manner for ditching at sea (the Honolulu surfing caper was still fresh in his mind) the actual occurrence would be dicey at best. Assuming the orbiter didn't break apart in the shock of bouncing across waves at over two hundred miles an hour, it would probably stay afloat no more than twenty minutes. The tail, with its three heavy main engines, would sink first. The astronauts would have to blow out the overhead windows with explosives; if by then water did not pour into the upright orbiter, they would sail away standing in insubstantial individual rubber rafts that resemble buckets and that are painted in colors supposedly invisible to sharks. Though in a real ocean ditching the astronauts' chances wouldn't be very good, this was, after all, still the simulator, and the astronauts' spirits remained high. Sally said they could have a great splashdown party, and Shannon asked the astronauts if they saw any unidentified flying objects in the Bermuda Triangle.

As they reset their switches, Browder and Williams discussed what they would do next. They had the same checklists and cue cards as the astronauts, and the idea that afternoon was to work through as many of the colored pathways as they could; they elected simply to move to the next one. It happened to be what to do when one engine conked out at Maximum Q, the moment of greatest aerodynamic pressure (and hence of the greatest stress and vibration) one minute after launch, while the boosters were still pounding away. At Max Q, the orbiter's three main engines are throttled back from full thrust to seventy percent of thrust, to ease the shaking, and then accelerated to full thrust once more. Sometimes full thrust cannot be regained, a situation astronauts refer to as being "stuck in the bucket," and that is what happened now. With only two main engines working at full thrust, they could do an RTLS once the boosters dropped away. This time they made it back to the Cape, without any more crashes.

Crippen seemed to be very much in command, and (except for an occasional one-liner from Sally) was doing most of the talking. Browder felt this was normal; he said that on most crews the commander does most of the talking, and this is particularly true during ascents and descents, when he would have the most to do. I missed some of the free-wheeling chatter of the rest of the crew from earlier days, and said I wondered if there might also be an element of everyone shaping up and buckling down when their boss returned. Browder nodded. "There is an element of that, and Crippen really *is* their boss, and people do tend to be a little more playful when the boss is away," he said. "Not that Crippen is the least bit a tyrant. He is in fact quite liberal in his management compared to some commanders." Perhaps a more important reason the crew was quieter, Browder thought, was that the astronauts each now had their own proper jobs to concentrate on, instead of taking turns at Crippen's, McBride's, or Ride's; this tended to isolate them a bit from each other. Another reason might have been that they were paying more attention to the procedures and checklists. Not long after Crippen had rejoined the crew, at one of the weekly status meetings, Browder had made an issue of the fact that they hadn't mastered the procedures and weren't always following them properly, and they had agreed that they were a little weak in that area. "It was just a word to the wise," Browder told me, "and there has been a vast improvement. Now one or another of them might miss a note in a checklist margin, but they would never miss a whole block of things because they were trying to do it from memory. And Crippen being there helps." Clearly the return of Crippen had resulted in a general pulling up of socks.

Ted turned away to talk to Steve Williams about the next abort. "What'll we do this time?" he asked.

"Surprise us!" said Crippen. Williams's microphone had been left on.

Williams turned off his mike and said, "Shall we go to the next block in the procedures?"

"No," Ted said. "Let's jump around and really surprise them." They decided on a late RTLS: one engine would fail almost four minutes into the flight, close to the borderline where they would be unable to make it back to the Cape. When it was clear they would be doing an RTLS, Sally hauled out the appropriate checklist card. "We'll be back just in time for the postlaunch parties," she said, starting down the colored track. One of the first tasks on the checklist is to dump the maneuvering system fuel—as with a conventional landing, it was best to get rid of it, not only to balance the orbiter, but to prevent an explosion if the orbiter landed hard. As Jon set about doing this, Dave reminded him unhelpfully, "Just make sure you're not dumping the thruster fuel overboard." Jon, of course, was now well beyond such a mistake, and Dave knew it. In fact, Jon was ready for ringing some changes on the old fuel-dump problem: when he flicked the switch for the dump, a valve leading to the vent wouldn't open—a malf improvised by Andy Foster, who had totally gotten over any generous impulses he may have started with. Jon quickly figured another route for the fuel to take, one that caused the hydrazine to be dumped out through the maneuvering system rockets themselves. Ted and Randy Barckholtz were pleased. They were a little less pleased a moment later, though, when (through no fault of Jon's) the orbiter began to go into an uncontrollable roll—something that had started when Williams failed a second engine. Crippen recommended a freeze. "One nice thing about contingency sims is that you can sure do a lot of them in one session," Sally said, as she began resetting switches for the next launch. They would get twelve launches in that afternoon.

Next time, they lost all three engines in the bucket (or, as Jon said, "in a deep well") and headed once more for the Bermuda Triangle. The time after that, they had one engine down and successfully did an RTLS. As they came in sight of the runway, Shannon asked Ted if she could do something to the computers. "Do it now," he said, and she did. Crippen solved the problem instantly. She pursed her lips. "Crip can psych out and solve a computer problem before the others even realize there is one," she said. "The highlight of my day is if I can kill Crip. Dave and Kathy have picked up a lot, but Crip's been working on the orbiter's data processing system from Day 1. There are times that I've screwed up and killed Crip—things that were my fault, such as trajectory offsets, when I've put the spacecraft too far off course to get back. He's had to dump it in the sea, but it's my mistake." I asked her if she had ever "killed" Crip legitimately. "Once I failed one of the four general purpose computers so that he had to engage the backup, and then I put in a nit—a transient failure—that caused it to disengage," she said. "Crip stood there not knowing what had happened. I had to tell him to reengage the backup; otherwise he'd have died."

It was a memory worth treasuring, because it didn't happen very often. Not only is Crippen a highly skillful astronaut, especially with computers, but he is also a highly competitive person. He once said that he liked Browder's comparison of training in the simulators with chess, and that the competitiveness was an essential part of both the game and the training. "You always try to stay ahead of the game— and in a sense it *is* a game," he said. "Sometimes, you even try to psych out your opponent, which may not seem like a real space flight, and if it happens too much, it may mean the crew and the team have gotten to know each other too well. But it is also realistic. In space or on the ground, if there's a malf, I always try to think what else can go wrong. And so does the team. And we all try to anticipate each other's moves. Even though they are moves in a game, they could really happen. So teaching us to stay ahead of the game is the most important thing sims can do, because we're all trying to psych out what might really happen." Crippen was better at doing this than anyone else, and like any person who is skillful, Crippen got pleasure from exercising his skill. The instructors were very much in awe of him. However big a pain in the neck Crippen could be in the simulators, one of the instructors told me that, if he or she ever got a chance to fly in space and could choose his or her commander, Crippen is the one he or she would pick.

It looked as if it was going to be a routine landing, something that clearly didn't please Browder. When Randy Barckholtz left to go to a medical appointment, Ted took his seat and picked up his light pen. When the orbiter landed, it mysteriously went off the runway. "I killed an auxiliary power unit," Ted told the astronauts, delighted with himself. "It didn't affect the flaps, but it did affect the nose wheel. I didn't tell you that Randy had left and I had taken his place." Evidently, before Ted had become a Team Lead, he had been as mischievous with a light pen as anyone, and doubtless had had to be held firmly in check by a tough Team Lead. Clearly in his new post he missed having his hands on the malfs. There seemed to be a conflict in Ted between mischief and restraint, and this clearly contributed to making him a good Team Lead.

"Let's kill all three engines fairly late," Steve said, as they reset their switches, and Ted concurred. One engine got stuck in the bucket, but nobody in the cockpit seemed very alarmed. In fact, somebody in the cockpit yawned.

"That was a Crippen yawn," Shannon said.

"It was McBride," Ted said.

They argued over the fine identifying characteristics of the yawn, without resolving the problem. Whatever the case, it was evident that a plethora of catastrophes could be soporific. Ted and Shannon felt challenged.

"What do you want to do, now, Stephen?" Ted asked, after the astronauts had once again landed in the drink.

"We've done every block on the front page. Let's go to the RTLS page," Williams said.

Shannon begged them to knock out all three main engines at liftoff—something that might wake them up in the cockpit. It would also demonstrate that without any help from the main engines, the two powerful solid-fuel boosters alone could lift the shuttle off the pad and get it far enough along its trajectory to land in the ocean.

At the moment of lift-off, Crippen asked, in a bored tone of voice, "What could possibly go wrong now?"

"Lift-off—and all engines are down," Ted, acting as capsule communicator, advised.

"If only they could lose a computer now," Shannon said.

"Do it, do it!" said Steve, but when she raised her light pen, he thought better of it and said, "No, no!"

Browder said, "She'll do it! And if she does, I'm taking out a fuel cell."

She did, and he did.

The spacecraft started wobbling up and down and sideways, and soon it was spinning. The motion-based simulator tilted farther and farther onto its side, and the horizon kept appearing, disappearing, and reappearing in the windows at an ever-increasing rate. "Oh, oh, oh!" said the astronauts, every time the horizon went by. Dave said it might help if they leaned the other way. One of the astronauts was whistling into his or her mike, making a sound like the wind.

On the way out afterward, Crippen said to me, "I hope we don't have to do any of those hairy things. Some day, someone will have to. And when they do, I think the orbiter will perform better than the simulator." With respect to the number of things that had gone wrong, both scripted and unscripted, he added, "If flights resembled sims, nobody in their right mind would ever go on one."

Clearly the astronauts' puppy days were not entirely over, and very likely never would be. Although the return of Crippen had caused a certain amount of shaping up within the crew, the crew seemed to be drawing Crippen into its own way of doing things. Not that Crippen needed much drawing in, for his crews always seemed to get along well; and to the extent that Sally Ride was a common denominator, and representative of Crippen's taste in crew members, they would always have fun—as Crippen wanted them to. Although the crew, in Ted Browder's opinion, had not yet crystallized, in the way John Young had meant, it was already very much a *crew,* and this fact was made clear by their initial reaction at having to take on two new members. As had been previously rumored, first one outsider, and then a second, had been added to their mission as payload specialists—passengers with special scientific projects. They wouldn't be crew members exactly; but then, there they *were* (or would be), getting in the way and taking up room in the cabin. How would Lindbergh have felt if he had been asked to accommodate two passengers in the back of his cockpit? Ted told me that, if the crew had been consulted, they would doubtless have voted five to zero not to have them at this late date. "I remember sitting around with them at a Happy Hour, all of us complaining," he told me; typically, Ted and the team empathized so fully with the crew that they had their noses badly out of joint, too.

Among other things, new passengers would screw up the crew's logo: a kind of apron would have to be appended below the circle, with their names and sex. Both were male, further unbalancing things. The two new guys (whom the crew had not yet officially met) were Marc Garneau, a member of the Canadian astronaut corps (something of a euphemism in the same sense as the Texas Navy), and Paul Scully-Power, an Australian-born oceanographer employed by the U.S. Navy who would be observing the seas. Not only was the crew clearly a crew, but—not unnaturally—it felt possessive about its mission, its spacecraft, and its windowspace, a scarce resource. "Seven people bothers me a little," Dave, perhaps the crew's most outspoken member, said. "Five may be a bit big for company, but seven's a real crowd. What we'll do with seven people all up and about at the same time, I don't know. And there's a lot of concern about whether the waste-management

system—the toilet—will hold up." The toilet had given trouble on
every mission so far. Dave evidently thought that I might think he was
complaining, for he concluded with a bright smile, "But we'll make it,
all right." The problem wouldn't come up for a while, because the
payload specialists wouldn't arrive for training until a few months before the launch.

*T*he crew might have been possessive about the mission, but Crippen
was possessive about the crew—or, at least, very protective of it, as a
good commander should be. He turned down a lot of requests from
me—that I be allowed to fly in a T-38 with Jon McBride; that I watch
Dave and Kathy inspect their spacesuits and EVA equipment in what
was called a "bench test"; that I attend certain meetings. With three
months to go until launch, Crippen's resolve that nothing interfere
with the crew's and the mission's safety tightened. Not only a crew
member, but an instructor, or even a planner, whose attention was distracted by a reporter's question, or by his mere presence, might miss
something that could be critical later. And the later in training, when
the instruction became more specific, the more important this could be.
    I learned the extent of Crippen's concern when I attended a flight-
readiness review of the orbital refueling system, which, unlike the
mock-up in the weightless environment training facility, would be carrying hydrazine fuel. Several times—by remote operation from the
rear console on the flight deck—it would be made to flow from one
tank to another, through internal piping, to prove that in weightlessness such emptyings and fillings could be done; and after Kathy and
Dave had done their EVA, it would be forced through the connection
they had made. To Crippen, it sounded like juggling nitroglycerin.
Hydrazine is explosive, toxic stuff, and Crippen didn't like the idea of
Dave and Kathy working in close proximity to it any more than he
would have liked sending them on an EVA in a payload bay full of
rattlesnakes.
    The flight-readiness review was held in a big meeting room on the
third floor of Building 4. There were perhaps fifty people there: technicians who had worked on the refueling and on the EVA tools, instructors from the simulators and the WET-F, and of course the astronauts.
Crippen, as commander, gave the signal to start—a thumbs-up sign.
As commander of a mission, Crippen had enormous power. If not in
fact responsible for everything to do with the flight, he might as well

have been; he had to be listened to on every aspect of its formation. Not only did the instructors have to defer to his definition of the crew's training needs, but the payload representatives had to defer to his objections and wishes. Though the building of a mission is a joint endeavor involving hundreds of people, including the entire chain of command of NASA and a number of its divisions and branches, a spaceship commander has the ancient prerogatives of a commander of a sailing ship who is ultimately responsible for the safety of the cargo as well as of his crew; and Crippen, being a navy captain, had a strong sense of this responsibility. Given his experience and prestige at NASA, it is safe to say that anything he seriously disapproved of wouldn't get aboard the ship. The more he heard about the hydrazine, the less he liked it—it could explode at temperatures of over 230 degrees, a temperature easily reached in the intense sunlight of space. One concern was that in space, in the shade, hydrazine in a pipe might freeze and contract, leaving a space; then, in sunlight, hydrazine would flow into the space, overpressurize the line when the frozen hydrazine melted, and rupture the line. Short of an explosion, the hydrazine was so toxic that if either Dave or Kathy got any on their spacesuits, once they got back inside the airlock and removed their helmets, the fumes could be lethal. It was explained that in the event of a spill, there was an elaborate set of precautions: the spacecraft would be turned so that the payload bay was toward the sun, and the astronauts would try to bake the hydrazine out of their suits. Then, in the airlock, before they took off their helmets, they would scrub down their suits with towels and detergents; in case they were running low on oxygen, they could plug their suits into the orbiter's own oxygen supply. Before they removed their helmets, they would have to seal the towels in airtight bags, purge the atmosphere in the airlock, and then pipe in fresh air. None of this made Crippen feel any easier.

The first speaker at the meeting was Harold E. Benson, the chief administrator for the refueling system, who explained that once the demonstration was successful, many customers would doubtless plan on using the space shuttle for refueling, thus providing the shuttle with much-needed business. Satellite manufacturers in the future would design satellites with quick-connects and -disconnects, to make them easier to service. The chief potential customer—the one with the most thirsty satellites already circling the sky—is the Department of Defense. As Benson talked, he waxed more and more lyrical. "So this general comes along and says, 'How about a demonstration?' "

he said. "The Department of Defense builds relatively simple satellites but lots of them, so in-flight servicing may be a big boon to them. They might need fewer satellites—they might need only six new satellites a year instead of nine. So the general came along and said, 'Let's develop a demonstration that would lead to refueling Landsat in 1986.' He said, 'Hurry up—I want you to do it faster and cheaper. I want to see an end-to-end demonstration, using crewmen and hydrazine.' " Crippen, whose face, Browder had told me, was the most expressive in the astronaut corps, frowned and pursed his lips.

When Benson was through, Crippen asked, "What do we gain by using hydrazine instead of water for a demonstration?"

Benson replied, "We would be demonstrating safe operating procedures, and we couldn't do that with water."

Crippen let out a deep breath and looked incredulous.

"We truly believe that all the obstacles have been overcome," Benson went on. "If we use water now, we would lose the respect of all the potential users. And using water would hurt the morale of the people who've been working on this job."

"It sure would improve mine, though," Crippen said, to general laughter. During the rest of the meeting, Crippen asked a number of questions, all directed at the safety of the operation; he appeared to have very little interest in any other aspect. He asked if they had done a shake test to make sure the refueling system would withstand the vibrations of launch and Max Q. Benson said, "No. But the refueling system is designed for stiffness, and therefore we can get by without vibration tests." Crippen looked unhappy. He asked about inspection techniques for welds in the tanks and pipes. He asked if certain valves had been certified. He asked if, for any reason, he had to bring the orbiter home early (abort the mission) in the middle of a hydrazine transfer, it would matter that there was pressure in the refueling system tanks and tubing. He was told there was no problem; and if, by some chance, the pressure rose above the safety point, the astronauts at the controls at the rear crew station would still have two minutes to respond. Crippen asked if there was a delay in their return requiring them to stay in orbit, which might cause the tanks to overheat, how long they would have to get the refueling system off the craft after landing. At the end of the session, Crippen still looked dubious, but he said, "You seem to have come up with a good system." He added, "I wish the ground could watch the system better, monitor it. The more eyeballs we have watching the system together, the better. We've got enough hydrazine back there to take off the rear end of the vehicle. I want to make sure there's nothing back there we don't know about. Let me talk a bit about hydrazine versus water. Hydrazine is dangerous stuff. If I'd asked to run a test of the refueling system, with hydrazine instead of water, while it was in the orbiter at the Cape, I know they wouldn't have let me do it, because there would be all kinds of safety regulations against it. But everyone seems to think it's O.K. for us to do it in space." Crippen not only was economical with words, but had a way of marshaling an argument that a litigator would envy.

McBride told me later he didn't think Crippen was all that serious about wanting to change the water, he just wanted to shake up the engineers so that they would make sure the system was as safe as possible. And Leestma said to me, "The ship is his responsibility, and that's where he's coming from. He's gradually leaning more toward hydrazine as he's learning more about the system. That's basically what the rest of us have done over the last six months or so." Over the last several months, while Crippen had been occupied with 41-C, the rest of the crew had become familiar with the refueling system and gotten comfortable with the procedures. It was one of the few areas where Crippen had to catch up. The rest of the crew had even participated in some of the design decisions—for example, when the engineers had wanted an automatic system for controlling temperatures and pressures in the tanks, Dave and Kathy had argued for a manual system, on the grounds that no one yet knew all the parameters of hydrazine under different temperatures and pressures, and they had won. This sort of participation in the course of training and planning gave the astronauts confidence.

Of course, participation didn't always breed confidence; sometimes it worked the other way. The technicians who had been designing the EVA tools for Dave to open the refueling system valves and make the new connection between tanks had finished the final versions (the actual flight articles), and they had recently notified Dave that they were going to pack them up and send them to the Cape, for eventual stowage aboard the spacecraft. Dave, whose experience with the prototype tools in the WET-F had not always been confidence-inspiring, told them not to do that until he had seen the final versions and had tested them himself. Dave, I had noticed, was one of the more picky members of the crew, never accepting things at face value—a useful trait for an astronaut, for in flight everything has to work the first time. Dave had the impression that the engineers were not anxious for him to do a final inspection. He insisted, however, and they gave him a time—an awkward time, very early one morning, just before the plane to the Cape was scheduled to leave. When Dave showed up, he discovered that one of the tools he would be using in space didn't fit. "Remember the ball valve—the pipe with the rotating valve that we screw into the fuel pipe and work through it to open other valves down inside the pipe?" Dave said. "Well, they had added a component to it, making it three-fourths of an inch longer than it had been, and that meant that the tool I would be pushing down through it to open the servicing valve below wouldn't reach. And here they were about to bag it and put it on the spacecraft! When I got out on the EVA, I'd have found it didn't work!" The tools did not get to the Cape that day. When Dave told the rest of the crew, they were incredulous; they told the story over and over again, and it rapidly became a legend at the Space Center, where each flight generates its own collection of such tales.

Not only did the mission have its potentially dangerous aspects, so did the training. When I saw Jon McBride after the refueling system meeting, he told me that a week or so before, when he had been flying a T-38 jet trainer over the Gulf of Mexico, he had had what he called a "catastrophic" oil-pump failure that had resulted in the flame-out of one engine, and he had been afraid that the problem would cross over to the remaining engine. "I haven't been shaken up like that for a *long* time," he said. NASA requires that all pilot-type astronauts—commanders and pilots, rookies as well as veterans—maintain their proficiency by flying at least fifteen hours a month. To do this, they drive out to Ellington Air Force Base, within ten miles of the Space Center, where the space agency keeps a fleet of twenty-six T-38's (the training version of an air force fighter, the F-5) as well as a couple of larger Gulfstreams also used in training shuttle astronauts.

Because I would not be getting my T-38 ride with McBride, I drove out to Ellington myself. It is a wreck of a place, because the air force has largely abandoned it, and the hundreds of barracks and other military buildings are dilapidated. NASA has kept a couple of big hangars and office buildings in operation, and in one of them I met Roger Zweig, a senior flight instructor and safety officer. Zweig is a forthright, businesslike man in his mid-forties who had once applied to be an astronaut and had been turned down. He took me first into a hangar full of sleek T-38's, painted blue and white, which appeared to

A Gulfstream II, modified to duplicate the flight characteristics of an orbiter for exact simulations of descents and landings from space.

A formation of three of the needle-nosed T-38s that look as if they are penetrating the sound barrier even when they are sitting on the ground. Astronauts use these training aircraft to get to their various appointments in different parts of the country, to keep up their flight proficiency, and also to practice shuttle-type descents and landings.

have "Supersonic" written all over them. I imagined the sound barrier
with holes in it, just the right size for the needle-nosed little craft to
slip through. Zweig allowed me to sit in a T-38. There are two narrow
seats, one behind the other. For all its sleekness on the outside, inside
it felt like an old Piper Cub or even a Model A Ford: compared to the
orbiter's cockpit, the dials on the dashboard are simple, businesslike,
and few. These planes are antiques, Zweig told me—twenty-year-old
antiques that NASA would like to trade in for something newer, if it
had the money to do so.

The pilot astronauts often used the planes to get around the country on business (it all contributed to their required flight time), and
Jon sometimes took Sally, Dave, or Kathy on their missions to Ball
Brothers in Denver, the prime contractor for the earth-radiation budget satellite; to SPAR, the factory in Toronto which had made the manipulator arm; or to the Cape to check on other equipment. But the
main thing they were supposed to be doing with the T-38's was getting
the feel for what it is like to land the orbiter. The trouble with the zippy
little craft was that the space shuttle, which they were meant to simulate, is anything but sleek and needle-nosed; it is a chunky craft that
looks like, and flies like, an overweight penguin that has been dropped
from a tall building. The T-38's look like, and fly like, swallows.
NASA's problem was how to turn a silk purse back into a sow's ear (or a
swallow into a penguin)—that is, to give the T-38's the aerodynamic
characteristics of the orbiter, which are about zero. It has a lift-over-
drag ratio of four to one—that is to say, if it were thrown like a paper
airplane, it would travel forward only four feet for every one foot it
dropped, which is not good. Some gliders have a ratio as high as forty
to one. The orbiter's poor lift-over-drag ratio doesn't matter much in
the upper reaches of the atmosphere, but the nearer the landing field,
the more important it becomes. From an altitude of about ten thousand feet, the orbiter approaches the runway at an angle of nineteen
degrees, a path seven times as steep as the one commercial jets normally follow. And as it drops down the last portion of its trajectory, it is
traveling at three hundred miles an hour. At about two thousand feet—
just short of the end of the runway—the spacecraft, with split-second
timing, pulls out of this dive by flaring upward, and for just the right
length of time this maneuver converts speed to lift, so that the spacecraft flies almost horizontally over the runway; when its wheels touch
down, the craft is moving at about two hundred ten miles an hour. This
is faster than any other craft lands. And unlike any airplane, the or-

biter, in the event of trouble, can't be waved off to fly around again. To get the same aerodynamic characteristics (or lack of them) from the nimble little T-38's, astronauts practicing approaches and landings open the speed brakes as wide as possible and keep them open, and for the same reason they put the wheels down at the beginning of the descent. They build up to these steeper landings gradually, long before they are even assigned to a mission. Jon took to it easily from the start. He and other astronauts have picked up a lot of experience in their T-38's as what are known as "chase pilots"—rendezvousing with one or another of the orbiters returning from space and accompanying them, usually with an observer or photographer in the back seat, almost to the ground.

When a pilot astronaut is assigned to a mission, his flying time increases to perhaps once or twice a week; as the mission gets closer, he will go more often than that. Once, when I asked Jon what he had been doing on a training flight, he said, "I went over the Gulf, did stalls, and aerobatics." Once or twice a month, he would spend an afternoon doing shuttle landings, usually at one or another of the landing sites that might be used: Edwards in California, the Cape in Florida, or at White Sands, New Mexico, one of the backup landing sites. On a White Sands assignment, which took an hour and a half to reach from Ellington, he might leave after lunch—following perhaps a morning's training in the mission simulators—stop for fuel at El Paso, get in ten or twelve landings at White Sands, and be back at his house in Nassau Bay in time for dinner.

Astronauts training for a mission also use one of NASA's two Gulfstream II's, which are larger craft fitted out for simulating shuttle landings with far more exactness. Instead of relying on flaps and wheels to increase drag, the Gulfstream's jet engines have been modified so that they can reverse themselves in flight, as most jets do to brake themselves after landing. There is a computer aboard, so that the exact flight characteristics the instructor desires can be punched in, down to the actual weight of the orbiter the astronaut will be flying, which varies, according to the cargo, from 195,000 pounds to 230,000 pounds. These weights, and their effect on the center of gravity, will affect the way an orbiter comes in for a landing. The astronaut and the instructor sit side by side; the instructor, on the right, has the ordinary controls of a Gulfstream, but the astronaut, on the left—where the commander sits in the orbiter—has all the controls of the spacecraft, and they respond the same way. There are screens in the win-

The 41-G emblem, fraught with symbolism, as most crew emblems are. In the center is a gold astronaut pin, three rocket plumes meeting in a star and encircled by an orbit. The constellations are two of the ones the crew had to memorize from their star charts; the one on the right consists of five stars, symbolizing the original five crew members.

The stand-alone team's emblem, a parody of the crew's. The stars on the flag are emergency and warning lights to signal malfs and glitches, some of which are written out on the white stripes of the flag, as they would be in a training scenario. In the center are three light pens of the type used for inserting malfs in the simulators, surrounded by a curl of electric wiring. The number of stars in the constellation on the right has increased and is arranged in a 7, to accommodate the two payload specialists, added later.

dows so that they give the precise field of view of the orbiter. Normally, the instructor takes off and flies to an altitude of about thirty-five thousand feet, something the orbiter would not be expected to do. The astronaut flies the descent from that altitude, does his flare at two thousand feet, and flies in level over the landing strip, settling slowly down—but he doesn't land. When his eyes are thirty-two feet above the ground, the height they would be when the orbiter touches down, a green light flashes on the dashboard—the purpose, of course, is to give the astronaut the sense of a real shuttle landing. The Gulfstream's wheels are still about ten feet above the ground. At that point, the instructor takes over and flies back up to thirty-five thousand feet, for another landing. When the lesson is over, they remove the screens from the windows to restore the Gulfstream's wider view, and the instructor lands the Gulfstream with the Gulfstream's own controls. No risks are taken in the airplane. Any malfs, anomalies, glitches, or nits during landing are handled in the safety of the simulators. The Gulfstream and the simulators are complementary, in the same way as are the weightless environment training facility and the KC-135 for simulating weightlessness, or the manipulator-development facility and the mission simulators for simulating use of the arm. In the simulators, all sorts of problems and disasters can be practiced, so that the astronauts get the widest possible exposure. The simulators, however, are finally just simulators, not the real thing. According to Zweig, the astronauts tell him the Gulfstream training is the most valuable they get. "They say that publicly," Zweig told me. "And what may be more important, they say it to us privately."

On Friday, June 15, the crew stood in again for the 41-D crew on an integrated sim, doing ascents. Because the 41-G crew's own computer loads wouldn't be ready for another month or so, there was no way that they could do their own integrated sims, nor would Mission Control have the time for 41-G's integrated sims until it was the next up. Now that the crew had broken the ice on doing integrateds, they generally considered them easier than ordinary stand-alones. Sitting next to Browder during one, I found the integrateds less lively—with the entire Mission Operations Control Center plugged in, things seemed to progress in a slower, statelier fashion. The astronauts didn't seem to be making as many decisions because the ground was making them more often; and since the scripts were generated by the Simulation Supervisor and another, larger team of instructors, whose scenarios were geared more toward Mission Control than toward the astronauts, Browder's team generated no malfs and nits of their own, but simply used their light pens to enter malfs and nits generated by others. It was an open question whether those other instructors got as caught up in their scripts as Browder's team normally did.

The afternoon passed uneventfully. With Mission Control making the ground calls to the crew, Browder's team didn't have that chore to perform, though they occasionally chatted, in their role of instructors, with the crew. And Browder talked over his headset with the Simulation Supervisor, or Sim Sup, Mel Richmond, who with his team sat in a small room that was to one side of Mission Operations Control Room, over across the duck pond. Browder was full of ideas for embellishing Richmond's script, most of which Richmond turned down. (In this situation, Ted played something of the same role toward Richmond that Shannon played to him.) Shannon made the only mistake I had seen her make so far: the script called for a computer to fail ten seconds after lift-off, but she put the failure in ten seconds before lift-off, so that the launch sequence automatically stopped and the spaceship stayed on the pad. The Sim Sup (*Sup* is pronounced as in the first three letters of *supervisor,* which of course is what it is) asked Browder what the hell had happened, and Browder asked Shannon why she had put in the malf too soon. Shannon, opening her big round eyes very wide, said,"Me? I didn't do that! You must think I'm dumb or something." Ted asked Andy and then Randy, in exaggerated tones, whether either of them had failed a general purpose computer ten seconds before lift-off; of course, since the computers were not in their areas, both of them shook their heads in profound denial. Then Crippen came on the communications loop to tell Ted to tell Shannon to shape up. Ted told Shannon she would have to defend herself. "You brought the whole sim down!" he said, in forced umbrage. "Fifty guys in Mission Control are sitting on their hands! Thanks a lot!" Everyone—even Randy Barckholtz—was delighted to catch Shannon out.

Even though Browder's team would increasingly have to share the crew with the integrated instructors, the stand-alone training, of course, would continue; and even in the final couple of months, when the crew was doing 41-G's own dedicated integrated sims most of the time, the integrated instructors would never have the same intimate relation with the crew that the stand-alone team enjoyed—and would continue to enjoy—at its consoles in the mission simulators. The team was a sort of shadow crew. Ted showed me an emblem Shannon had designed for the team, which was a parody of the crew's own emblem: the flag in the upper half of the circle had caution-and-warning lights instead of stars. Their names arced around the bottom: Barckholtz, Browder, Foster, O'Roark, each followed by the appropriate male or female symbol. And in the center, instead of three rockets zooming upward toward a star, Shannon had drawn three light pens converging; instead of being held together by a golden circle representing an orbit, they were surrounded by the curling, stretchable wire that connected their headsets to their consoles. It was, I told Ted and Shannon, very nice.

The Measuring-Unit Dilemma and the Freeze-dried Computer

By the last week in July, the pace was quickening. There were just a little over two months until launch. When Sally and Dave went to the manipulator-development facility to hoist balloon mock-ups with the arm, the balloon was in the shape of their own satellite, the earth-radiation budget satellite, and no one else's—it looked like a car radiator with a row of bottles on the top. Similarly, the payload bay they looked down upon represented their own. And the computer loads for their own mission had arrived at the mission simulators. Never mind that the orbiter whose software was duplicated was the *Columbia*'s and that the visuals of the prime landing site were of Edwards Air Force Base, for the loads were already in production when the change was made to the *Challenger* and Kennedy; it was still *their load,* with their own trajectory highly inclined to the equator, and with their own payloads. Updated loads using the *Challenger* and Kennedy would come in a couple of weeks, they had been told. Despite the discrepancies, their arrival had been a major milestone, for Browder said to me, "Getting your own loads is almost like starting over again. Or maybe it's like jumping from college to graduate school."

Kathy Sullivan was perhaps the most pleased, for right up at the front of the cargo bay was the OSTA package, containing the shuttle-imaging radar antenna as well as three other experiments, the large-format camera, the measurement of air pollution from satellites experiment, and the feature identification and location experiment, all of which were chiefly her responsibility, and which she could now practice with. For the first time, she would be able to get on with the kind of work that had led her to be an astronaut in the first place, the operation of scientific instruments in space. In 1978, when she got her doctorate in geology from Dalhousie University, she had considered being an astronaut (her brother, who had applied but been turned down, told her that he thought a woman scientist would have a good chance), but instead she had gone on to be chief scientist aboard an oceanographic vessel. "I dealt with the bottom of oceans," she told me. "I wasn't sure what I'd get from one or two hundred miles up, or what NASA would get from me." But the longer she was at sea, and loving it, the more she thought about space. "I thought that being a mission specialist might be similar to being the chief scientist on a marine exploration vessel— keeping investigations going, despite storms and breakdowns," she said. "Of course, I enjoyed getting my own data, doing my own work. But my favorite was being at sea, keeping all the experiments running, whoever's they were. And NASA wanted this."

With the arrival of the loads, there had been a decided shifting of gears. Kathy in particular had the air of a mariner, long confined to land, stowing some favorite piece of equipment aboard ship and savoring the smell of the sea. She told me, "We're so busy that we can't really think about the space flight, but we know that we're getting close. Little things remind us. Often now we find an exercise is the last time we will do that particular thing before we go. And the flavor is definitely changing. The flight directors and the sim planners are considering what *we're* going to do, not what some other mission is doing. In the simulators, we've moved from doing generic skills of the orbiter, applicable to any mission, to simulating our own mission-specific flight plan. It all becomes *ours*. We're still doing ascents and entries, but we're also spending more time on what will keep us up there and busy for eight days—our own in-orbit activities. For the first time, I have to think, 'What do *I* really need to remember? What do *I* need?' It's like getting close to the time to start on a camping trip or a cruise: we're moving from questions like, 'Do I remember how to run the gas stove?' to really specific stuff like, 'Where are my long woolen socks?' All of a sudden, it's become real." And Jon McBride said, "It's getting so close, I can almost *feel* it."

The astronauts now were next-to-next up, number two in line. This had not happened in quite the orderly manner that had been planned: 41-D should have been launched late in June, followed by 41-E on September 1, and 41-G on October 1. In fact, 41-D had not gotten off on time. Because of a failure of an actuator in the main fuel valve in one of the *Discovery*'s main engines, all three had shut down seconds before they were to ignite, and about three seconds before the boosters were supposed to blast off. Had the boosters ignited, there would have been no holding the *Discovery* back; it would have taken off with no main engines firing and landed in the drink, as per the scenario Shannon had been so anxious for the 41-G astronauts to experience in June. The 41-D crew might have been the first to use the stand-up floating buckets painted the colors invisible to sharks. Fortunately, nothing happened, except for the flames licking around the rocket engines that had so surprised McBride. Before the launch, the 41-G crew, minus Crippen, had flown down to the Cape together in the Gulfstream—Sally, of course, had had the choice of sitting with her crew or with Steve Hawley's family. Crippen, who had been in Washington and watched the aborted launch from his hotel room on television, told me later, "It was a disquieting episode, but not an alarming one. If anything, it was reassuring. It proved that the whole stopping sequence for the rockets worked perfectly—and we never would have been able to test that. Nobody panicked. Nobody did anything foolish like asking the crew to go down the slide wires to the fireproof bunker, an escape we have never had to use. All in all, I was pleased." Later, he was a little less pleased, because, in rescheduling 41-D, one option had been to scrap 41-G. This notion, Crippen had found to his dismay, had support at NASA headquarters because almost all the 41-G payloads were NASA's own, and therefore its cancellation would have the least adverse impact on customers NASA was desperately anxious not to lose. One factor that very likely helped preserve 41-G was the orbital refueling system experiment that Crippen had been so suspicious of, for NASA management was very much aware of the air force general's demand for a demonstration that could lead to a lucrative business refueling Department of Defense satellites. The result was that 41-E was canceled, 41-D was now scheduled for September, and some of the more important payloads of 41-E were added to 41-D's manifest, as well as to other flights later on. The 41-E astronauts—who were doubtless at least as upset by their loss as the 41-G astronauts would have been because they were even farther along in their preparations—would be reassigned to another mission. Consequently, the abort and the rescheduling had no adverse effect on 41-G, which was now scheduled for October 1. It was still second in line, as it would have been had 41-D got off on time. Yet the uncertainty (nothing at NASA is ever sure until after it is done) added an element of suspense to the crew's anticipation. Kathy said to me, "Part of the excitement is not being sure of the exact date; we know it can slip. So it's not like the excitement of looking forward, say, to the beginning of summer vacation, which can be pinpointed. But we know we're sneaking up on it."

*T*hrough all this, William Holmberg and the other composers of the crew activity plan had once again put the 41-G timeline on the back burner until things settled down, and then brought it forward again. There was a Flight Operations Review meeting the last week in July, at which a number of conflicts were resolved. As had been the case with the orbital refueling system meeting, which had been run by Harold Benson, this meeting was not chaired by Crippen but by Holmberg; as on the earlier occasion, Crippen nonetheless seemed to dominate the proceedings, from a chair inconspicuously placed off to one side. This time, though, he shared his obvious authority with a stocky man with a walrus moustache who was sitting next to him—John Cox, the lead flight director for 41-G.

Like Crippen, Cox had been involved with the 41-C mission (he had been in charge of the team of flight controllers that handled the two EVAs) and hence he had only recently become involved full time with 41-G. The lead flight director, who is in overall charge of the flight controllers running a mission from the Mission Operations Control Center, exerts almost as great an influence on the evolution of a mission—planning as well as training—as the commander. This becomes more true in the latter part of the training, as the flight controllers become increasingly involved. His role in the mission so far had been very much behind the scenes, working on the ways and means for carrying out the maneuvers called for in the crew activity plan and developing what are called the "mission rules" governing such matters as under what circumstances the satellite might or might not be launched, or whether the scientific instruments in the payload bay could operate during the EVA, as well as stipulating a variety of other dos and don'ts for operating the payloads. This sort of analysis often involved computer simulations. Cox knew at least as much about everything that was going on in the mission as anyone, including Crippen, and shared with him many of the final decisions. The two met increasingly often. If the space flight were a theatrical production—and in many ways a space flight is one—then Crippen and the crew would be the stars, and Cox the director. He has known Crippen well, ever since the early seventies, when, as Cox recalls, they "competed together in various tug-of-war contests and beer-drinking events." He knows the others well, too. Jon, Sally, and Dave have all served as capsule communicators on other missions he has directed.

The plan, near its final form, now postulated an eight-day mission with launch at dawn of Day 1, which was October 1, and landing mid-morning of Day 9, October 9. The EVA was scheduled for Day 5—late enough so that the astronauts would have adjusted to weightlessness and recovered from any nausea (which affects about half of all astronauts during their first few days in space) but with enough leeway before landing so that it could slip a day or two, if it had to. The German SPARKS, which had been added to 41-G's manifest at the last minute, had been removed. As more and more items became firm, it was harder to introduce changes. There were still some open questions, though. At the meeting, the earth-radiation budget satellite people from Goddard said they were anxious to get a photograph of the satellite after it was deployed, to document its condition. When should this be done? Crippen drummed his fingers on a table without replying. Cox made some jottings on a piece of paper and passed it to him. In the note, he pointed out that when Crippen pulled away from the satellite, he would do so by slowing the orbiter slightly, so that it moved backward from the satellite and dropped into a slightly lower orbit, which would cause it to speed up and bring it by the satellite again, though underneath; wouldn't that be a good moment for the photography? Crippen nodded his agreement, and it was decided. The shuttle-imaging radar antenna was another problem; when it was opened, in order to transmit and receive radar signals bounced to and from various target areas on the earth below, it extended to thirty-five feet by opening two panels on either side of a central panel to form what looked like a gigantic billboard. The antenna was so flimsy that its designers feared it would be damaged if it was left open during an orbital maneuvering system burn. And there would be two pairs of burns in the days after the spacecraft reached its original high orbit one hundred ninety miles up: one pair to bring it down to its middle orbit of one hundred forty-eight miles and to circularize it there, and another pair to bring it down to its lowest orbit of one hundred twenty-one miles and circularize it. The lowest orbit, of course, was best for gathering radar data, but the scientists didn't want to miss the opportunity to gather data at the two higher orbits. The question therefore was, Should the radar antenna be deployed before reaching its lowest orbit? If there was a malfunction so that it couldn't shut automatically, it might break. Kathy and Dave were dispatched to the WET-F to see whether they could close the two side panels on an emergency EVA, and the answer was that they could. Consequently, the antenna would be opened at both higher orbits. There was also a lot of horse trading to be done between the radar people, many of whose targets were to one side or the other of the orbiter's ground path, requiring that the orbiter's attitude be shifted to one side or another of the vertical, and the proprietors of the other OSTA experiments, the large-format camera and the feature identification and air pollution experiments, whose instruments had to look straight down. According to Holmberg, the radar was what he called "the big driver" of the orbiter's attitude—most of the time, the orbiter would be flying with its nose forward along its trajectory and the payload bay straight down, so it would be relatively easy to shift this way and that to please the radar experimenters. Every time they did that, though, the other scientists would lose data. In addition, attitude changes had to be included to oblige Canadian scientists so that their payload specialist, Marc Garneau, could photograph auroras; the overhead window, used for photography, would have to be pointed toward the northern or southern polar regions when they were dark. On those occasions, the radar people would lose out, too. Crippen said he would do this for the Canadians no more than four times. This was not out of resentment of the payload specialists but because there were already a lot of attitude changes, most of them for the radar, that had already been approved; there would be one hundred thirty-five in all. Crippen, McBride (who would be doing most of them), and Holmberg didn't want any more; they agreed that the number was already so great that they would clutter up the timeline, where they were scattered throughout in such a way that it would be hard to find the next few at a glance. Accordingly, they agreed to remove them from the chronological main body of the timeline and include them all together in an appendix at the end; that way, McBride could work down the list more easily. Crippen had endorsed the rest of the crew's efforts to have routine chores lumped together at the start of each day's schedule, allowing them to do them when they could. Uncluttering the timeline was something he was entirely in favor of. Among other reasons, doing so gave the crew greater independence.

At the simulators, the crew had begun using the new computer loads, the ones that had been made for their own mission, despite their differences from the present plan. That day—Monday, July 23—the crew and the team would practice an ascent all the way to orbit, followed by activities in orbit; it would in fact be one continuous simulation of the first four hours of flight, with no aborts. The longer a single simulation lasted, the more complex it could be, because problems and their consequences could build up over a period of time in a manner they couldn't in, say, a series of ascent simulations, none of which might last more than fifteen minutes. Of course, both types of simulations would continue to be used, but there would be an increase now in simulating longer and longer stretches of time, building up to whole mission days and culminating, in the integrated simulations, with one that would last three days. This, of course, was the only way to get at the mission-specific complexities of operating their own payloads in orbit—and time in orbit does not come in neatly packaged small units, as ascents and descents and aborts do, but in long stretches.

As they reset their switches, the astronauts chatted together as usual. Sally Ride said she had just started jogging again; she had finally got the idea they would fly soon and she had better get in shape. Dave Leestma said he suspected the reason she had taken up jogging was that she was getting ready for the chicken dinner circuit as a spouse of a 41-D crew member—astronauts just back from space, and their families, often go on publicity tours, which include a number of banquets. In the control room, the team, too, was resetting switches. Shannon was away for the afternoon, and her place was taken by a computer expert from another team, Alan Fox. Shannon, though, had left the script she had helped create for him to follow. Soon after lift-off, he put a glitch into a multiplexer-demultiplexer, and MDM, which translates data between a computer and several of its strings. Jon McBride noticed that one of the affected strings controlled the right orbital maneuvering system engine, which would be soon needed—following main-engine cut-off eight and a half minutes after launch—to give the spacecraft its final push into orbit. He pointed this out to Crippen and suggested that he move the strings to other computers. Crippen liked the idea; he told Jon to go ahead. Ted Browder was delighted. "Jon has really gotten over his reticence with Crippen," he told me. "He got over it a week ago, when there were a lot of problems on his side of the cockpit. Jon just leapt into one problem after another and told Crip what he was going to do, and then went ahead and did it. He's really coming into his own." And Ted was just as pleased with the three mission specialists. "They're like kids helping mom and pop around the house," he said. "They've fallen into knowing what sort of things Crip and Jon don't want to do. They're so aggressive in wanting to help, it amazes me! So many mission specialists want to stay out of the way of the commander and the pilot; they don't want to press the wrong switch. But their eagerness, of course, is all within the bounds of a homogeneous crew." He attributed this to the experience they had gained while Crippen was away, and which was proving useful now that he had returned.

Jon, however, had inadvertently managed to foul up Fox, who was working his way down Shannon's script. Further on, Shannon had planned what is known as an inertial-measurement dilemma—two units would disagree on the attitude and velocity information they were providing, and the crew would have to decide which of the two to believe. For the problem to work, of course, the third unit had to be down, and Fox had failed it earlier. By restringing, McBride had brought the failed unit back on line. "Shannon didn't write the script well," Fox complained.

Browder told him to forget the dilemma problem since Shannon obviously had not realized that the restringing would bring the unit back. Fox, Ted told me, came from one of the teams that rarely use scripts but rather wing their scenarios, and he was showing his distaste for scripts because they could get you into trouble. In Ted's view, of course, no-script teams were clearly analogous to hot-dog pilots, and he wasn't pleased that Fox was picking holes in Shannon's scenario. After this skirmish, they chatted amiably a bit, exchanging gossip about their teams: an instructor who had been with a very conservative team that never departed from its scenarios had recently joined Fox's group, and, according to Fox, had completely changed his way of doing things—he had been liberated. Ted looked skeptical at the news of this conversion. Before he got back to work, though, Ted told me it was good for each team to know what the others were up to.

It turned out that the crew didn't need the orbital maneuvering system engine to push the spacecraft into orbit, because in the meantime Andy had inserted a malf so that one of the three main engines was stuck on and did not shut down at main-engine cut-off: it continued to burn, providing enough thrust to attain orbital velocity. This "overspeed," as Ted called it, was the opposite of the problem of underspeeds caused by getting stuck in the bucket. Meanwhile, Andy had put a major leak in the nitrogen system for one of the maneuvering system engines—nitrogen is used to pressurize the hydrazine fuel, and without it the engine wouldn't burn properly. Andy, it seemed, had overcome his scruples against being unkind to the crew. Even though the overspeed had removed the need for the maneuvering system engine's firing following main-engine cut-off—called the OMS 1 burn—it would be needed some forty-five minutes later for the OMS 2 burn, which would circularize the spacecraft's trajectory when it reached the higher orbit. In the meantime, Fox had killed one of the four general purpose computers altogether—strings and all—and as a result the data about the nitrogen leak wouldn't show up on the astronauts' screens. The screens were down. It was about eight minutes before the burn when McBride, who was checking out the engines, found he couldn't call up the data he needed. The ground would still be getting it, and he wondered out loud if the spacecraft was in contact with Mission Control.

"There's one easy way to find out," Ted said to his team.

Jon thought of it, too. "Houston, *Challenger*," he said.

"*Challenger*, Houston," Ted answered. They were in contact. Ted went on to tell Jon that there was no pressure left in one of the maneuvering system nitrogen tanks, so he would have to do a one-engine burn, which meant burning the good maneuvering system engine twice as long. Ted monitored the burn carefully. He wanted to see if Jon would remember to cross-feed fuel from the tanks on the other side halfway through the burn, in order to keep the orbiter balanced. He did, and Ted looked relieved. "Jon's coming along fine," he said. "Our team has meetings from time to time, and the last input from Randy was that Jon is 'there' in electrical systems, but Andy still feels that Jon has a little way to go with the maneuvering and thruster systems. His main problem is in identifying and isolating leaks, especially when both those systems are involved at the same time. It seems to be the last area where Jon isn't as intuitive as he is in all the other areas. It's not that he can't solve maneuvering and thruster system leaks; he *can*, but he has to pause and concentrate his forces. And if anything else goes wrong then, it'll throw him off. Almost everything else he does is instinctive now. In this area, I know he'll get there soon."

In orbit, the astronauts had to set up for orbital activities, and this meant putting away the folding seats Sally, Dave, and Kathy had sat in during the first part of the ascent, unpacking the things that they would need, and generally getting things ready. Dave announced that he would start working on the middeck, and Kathy said she would get to work on the rear of the flight deck. "Of course they know what to do without saying so," Ted said to me. "They're just saying, 'O.K., we're ready to go.' They're simply acknowledging to each other it's time to get on with these chores." They sounded very close to being crystallized.

Randy wanted to put in a malf that would give them difficulty in opening the cargo bay doors, the pair of huge, sixty-foot-long panels that slowly separate like giant clamshells to reveal the cargo bay; inside each is one of the two big, silvery radiator panels that move on hinges up from its door. Through these the freon cooling system dispels the heat from the cabin and the electronics into space. The electronics were already hot from the activities of ascent; the doors had to be opened, and the radiators exposed to space, as soon as possible. Ted told Randy to make the malf an easy one; if it was too hard, either Kathy and Dave would have to go on an EVA into the cargo bay to open the doors, or else they would have to de-orbit and come home. Either way, it would foul up the sim. "The basic purpose of the sim is to give the crew a look at the in-orbit timeline—we're expanding the first two hours into four, to give them a good feel for it, so we don't want them to come home," he said. Randy simply popped a circuit breaker, which Kathy quickly recycled—that is, she flicked it off and on again.

While Kathy and Dave were getting the spacecraft squared away for being in orbit, Crippen and Sally were trying to see what they could do about the computer situation. When the third of the four primary computers—number 3—had died, Crippen had instantly brought a backup computer on line. This meant reassigning the roles of the various computers, or restringing. In the course of restringing, the fourth computer, which contained the stored information for entry (called the "freeze-dried" entry program), was given a different function, so that it could no longer carry out its freeze-dried duties. Alan Fox could hardly believe it. Having caught Shannon out, he was now about to catch Crippen out.

"They just *blew* it!" Fox said, as delighted at the prospect of "killing Crip" as Shannon had ever been.

Ted looked over Fox's shoulder. "That's normally Crip's area, so I'm surprised," he said. He told Fox, as he had told Randy, that, in the interests of the sim, he didn't want to have to abort and come home. A query from the ground to the spacecraft was definitely in order.

Fox called the *Challenger* and asked the commander what he planned to use as the freeze-dried computer.

"We have no freeze-dried computer," Crippen said. "We're using the mass memory as the backup instead." In addition to the five com-

puters, there are two magnetic-tape mass memories for bulk data stor-
age, and all the entry activities information had been stored in them,
in addition to computer number 4.

"Where in the procedure does it say that?" Fox asked.

"It's in the Computer Malf book," Crippen answered, citing section, page, and block number from memory.

"And will you sign today's training report for Alan?" Browder
said, clearly delighted that Crippen had taken a fall out of Fox. The
training report is what all the astronauts' teachers have to sign after
each lesson, to prove that each square in the syllabus has been covered. "I just want to make sure Alan gets credit for today's lesson." He
slapped Fox on the back, and said, "We all learn. We should tell Shannon about this; we all get bit the same way by Crip." Shannon would
probably be able to use some moral support in any future encounter
with Fox over the inertial-measurement unit problem. Browder felt a
little sorry for Fox, however, and Ted wasn't the sort of person to let
anyone suffer for very long. Ted slapped him on the back again and
confessed that he, too, had been bitten—he admitted that *he* had not
known that the mass memory contained the freeze-dried information,
either. "In the orbiter, there are so many things you can do to solve a
problem," he told me later. "Alan didn't realize that there was another
option open, different from the one that he was thinking of—as Crip
showed him in the procedures. I hadn't realized it either." The orbiter,
I was beginning to see as the lessons got more advanced, was like a
living organism in its complexity as well as in the relationship of its
computers to the rest of its systems.

Fox, however, was not easily mollified. He said he wanted to throw
a star-tracker problem at Crippen, but Browder wouldn't let him, because the astronauts were behind in their timeline now, and the object
of the sim was to keep them on it so that they would understand its
intricacies. "I want to get them back on the timeline by leaving them
alone for a bit," Ted said. He may also have felt that Fox had had his
shot at Crip.

Kathy, at the rear console, deployed at the aft port side of the cargo
bay an antenna for the Ku-band radio, a black graphite epoxy dish,
which would be used for transmitting data from the scientific instruments to the ground via the relay satellite. Next, following her procedures, she began turning off much of the caution-and-warning system. Many of its alarms were useful during ascent, but in orbit they
would be going off constantly, taking up the astronauts' time with

minor nits that the ground could monitor and control. Up in the front of the cockpit, Crippen and Sally were checking the electrical system, a task that occasionally required them to set off an alarm as part of a test. Suddenly one went off when it wasn't supposed to.

Crippen and Sally looked at each other, and then they looked back at Kathy. "Kathy, what are you doing?" Crippen asked.

Kathy told him she was inhibiting the alarm system.

"You set off a Caution and Warning, and I can't figure out what it is," Crippen said.

Kathy apologized; evidently she had set off an alarm by mistake. Crippen asked her to stop inhibiting the system until he and Sally were through with their tests.

Later, when I asked Ted whether he thought the team had crystallized yet, he shook his head, citing the example of Kathy's inhibiting the caution-and-warning system while Crippen and Sally were trying to use it. "If you define *crystallized* as the crew's being 100 percent able to anticipate, as each member being able to comprehend everyone else's duties, for all of them to understand Crippen and each other—to act as one person—they're not there yet. But they're very close."

Shannon O'Roark breezed in from wherever she had been—she was wearing a billowy pink blouse, black slacks, and high-heeled sandals. "I missed you guys," she said.

"You made a lot of mistakes in that script," said Alan Fox, who was evidently still burning from the Crippen freeze-dried episode.

"I didn't miss you guys," Shannon said, turning swiftly back out the door. Fox moved rapidly after her, and soon loud voices were heard in the corridor.

One reason that the crew was still some way from crystallizing, in Browder's opinion, was that there were a number of technical problems with the new loads, resulting in a series of sim crashes. New loads always had problems that had to be ironed out, and those particular ones were no exception. In addition to being for the *Columbia* instead of the *Challenger,* there were a number of nits, and these were particularly evident in the ascent and contingency abort loads, which seemed always to end in sim crashes. Today they were doing transatlantic aborts to various airports in Europe, but most of the time the astronauts landed in the drink or worse. On the first run, they lost

computer number 4, and Crippen told Jon to restring. Then they lost a
main engine. As they headed for Spain, Crippen reported he was getting a "big divergence in altitude"—that is, his altitude was several
thousand feet below where it should have been. "We're losing control,"
Crip said, asking for a freeze. "Something weird happened." The next
time, they made it all the way to Zaragoza, an American military field
in Spain, but weird things began happening so that Crippen had to
take manual control earlier than usual, and he demanded another
freeze. As the next one missed its abort landing field, Dave, who had
been at another appointment, popped into the control room, and when
he saw what was going on, he said, "I hate these LOCs"—LOC is short
for "loss of control." Ted said he didn't like them either, and Crippen
came on the loop to say that that was three LOCs out of three.

Browder was upset. He told me that out of the last eight missions,
including today's—that is to say, in all the crew's simulator sessions
since the new loads had arrived—six of them had had so many sim
crashes that they all had to walk out of the simulators and go home. To
add insult to injury, other crews who were using the 41-G loads (in the
same way that 41-G used to use the loads of crews ahead of it) had had
many fewer crashes and were getting much more training out of them.
"It's very serious," Browder said. "Now we really are behind." Crippen
was worried, too.

I asked Browder what he could do about the situation. "First, we
can rearrange the schedule, so that the most important lessons are up
front, and the less critical ones come later. That makes it easier to decide what we can cut out. We *do* tend to overbudget, so if a small percentage is missed, it won't jeopardize the mission. If we get to the point
where I see that we're losing some critical stuff, we'll start bumping
other crews behind us, who are working on less critical things—say,
like familiarization. The farther downstream they are from us, the
more we can bump them and use their time. As a last resort, we could
simply bump everyone behind us and just move in here until we got
the job done. And the last resort of all would be slipping the launch
date of the mission. NASA headquarters would be involved. However,
a mission has never been slipped for lack of training." He concluded,
"But we would never launch a crew that was untrained."

That week, the astronauts had a photography class in the one-G
trainer. There are normally about twenty-two cameras aboard the or-

biter, including several television cameras (four in the corners of the cargo bay and two on the arm), a movie camera, and three still cameras—one, an Aero-Linhof wide-angle camera that uses seventy-millimeter film and is good for out-the-window photos of the earth; one that uses thirty-five millimeter film and is used indoors; and a third, to be taken on the EVA, suitable for documenting the work on the orbital refueling system. Kathy would use this one. Teaching the astronauts how to stow and unstow all this photographic gear, and how to use it, was the job of Stephen Seus and a handful of others. Seus is a dapper, photogenic instructor, well tanned and frequently dressed in a blue blazer. Individual astronauts, and even crews, vary in their interest in photography and their proficiency at it. Crippen was very encouraging about photography and tried to get a lot of it done on his missions; Jon and Dave were interested in photography, too, though they weren't any great shakes at it. Seus urged the astronauts to take home with them thirty-five millimeter cameras such as they would have aboard, to practice with on their own families, or on anything else that might have what Seus called "good photo opportunities," and to bring him the film, which he would develop and critique. Jon and Dave had an unfortunate way of chopping off heads or feet. Kathy was perhaps the best photographer in the group; she had had experience before. As a geologist, she had expressed the greatest interest in the seventy-millimeter camera for photographing the earth. Sally seemed to have a rather relaxed attitude toward photography—in fact she had arrived very late at this morning's session.

In the one-G trainer, they had discussed a new piece of equipment: for the first time, NASA was sending up photographer's floodlights to be used with the TV and movie cameras for photography indoors. (There was a new sixteen-millimeter movie camera, an Airflex, which cost forty thousand dollars. It had a zoom lens, and one of the most difficult things Seus had to get across to the astronauts was not to zoom the lens while they were actually filming—an irresistible idea to any amateur photographer, but not to a professional, who normally sets the lens at the desired length ahead of time and leaves it there.) The crew had experimented setting up the lights in various parts of the cockpit, but they always seemed to be in the way of one or another of the astronauts. At length Jon, who is very accommodating, offered his own pilot's seat as a good place; he was not apt to be sitting there himself when pictures were being taken. Accordingly, that morning, they had folded up the pilot's seat and put the floodlight there. "That was a good input," Seus said to me. "We still have to get Crip's permission. He may not like folding up Jon's seat—he may want Jon to be able to get into it quickly, in the event of an emergency." After that, they had practiced taking seventy-millimeter photographs out the window with the Aero-Linhof wide-angle—they had to learn that because of its large field it was best to aim it straight down at the ground below, not obliquely at terrain to the side; and that in October, when they would be flying, large sections of Europe and North America tend to be cloudy, so perhaps they should focus on the Southern Hemisphere.

Different groups had requested certain specific pictures—the payloads people, for example, wanted documentation, not only of the satellite after it was released, but of the radar antenna, and of course of the refueling system during the EVA. The Public Affairs Office always submitted a long list of requests of the astronauts eating or sleeping or generally horsing around. Seus had asked the astronauts about their own personal preferences. Most of them wanted to photograph the earth or each other; Seus showed them how to cover the flashbulb with a handkerchief, so that there would be less glare in the interior candid photos. Jon was anxious, Seus said, to take some funny photos—comedy routines. Astronauts have sometimes been good at this —Jack Lousma, for example, the commander of the third mission, had been filmed doing push-ups, first with two arms, then with one arm, then with none (his toes were hooked under a console, and he raised and lowered himself with his ankles). Joe Allen, a mission specialist on the fifth mission, had caused a ball of orange juice floating in the air (all liquids form spheres in space) to disappear on camera, by sucking it up with a straw. According to Seus, the ideas for what he called "antic photography" were running out, and the astronauts on each mission had to strain themselves a little harder to think of any. There was a certain amount of repetition; Seus himself was getting tired of such shots as astronauts tossing their food in the air and catching it in their mouths, like seals at feeding time. (The most memorable such picture had been on Crippen's 41-C mission, after those astronauts had succeeded in catching the Solar Max with the arm, using a technique called a rotating grapple; James Van Hoften had demonstrated another sort of rotating grapple on camera when he had partially peeled a banana, rotated it so that the peels flapped around like the blades on a propeller, released it, and caught it in his mouth.) The 41-G crew had spent some time that morning discussing possible antics; one suggestion had been that an astronaut might levitate himself horizontally in the middle of the cabin and then see if he could blow himself around, using his mouth as a rocket. He would, of course, have to throw his head back so that the thrust would go through his center of gravity, as if he were regimbaling the maneuvering system engines for a single-engine burn. When the class broke up, Jon McBride asked to borrow a thirty-five millimeter camera; he planned to go flying in a T-38 that weekend, and he thought there might be some good photo opportunities.

*I*n addition to all these cameras, which belonged to NASA and were more or less standard equipment on all missions, there was one huge camera, the IMAX, that weighed about seventy-five pounds and that had a film magazine weighing an additional thirty pounds. It belonged to a Canadian client, the Threshold Corporation, which used its IMAX camera for making special large-sized films for exhibition in a few theaters that had special extra-large screens. The IMAX film called *Hail Columbia,* playing at an IMAX theater at the Cape, had been made on the ground some years before; it included a live launch, in which the main engines and the boosters seem to light up practically in the audience's laps. The film had impressed NASA management, who were convinced that flying an IMAX camera in space would be as dramatic a way as any of involving the public more closely with space flight, short of actually having the public in the cockpit. This, of course, is an argument that Crippen and most astronauts, who are generally opposed to taking nonastronauts into space but want their support, would embrace. NASA agreed to carry the camera aboard the 41-C, -D, and -G missions, to obtain footage the company proposed to turn into a new space filmed called *The Dream Is Alive.* Crippen, who would have the camera on two flights he commanded, was especially interested in the project; on 41-C, he had done most of the photography himself, though on 41-G he promised to let Jon McBride be the cameraman. The IMAX training was done by the head of the company, a soft-spoken Canadian, Graeme Ferguson, who had an intense belief in his product. The pictures already back from space had given him great hopes—the footage from the 41-C mission, which he had already seen, had surpassed all his expectations. In training, he was concentrating on teaching the 41-G crew the nuts and bolts of the camera, which are complicated. As this was to be an earth-watching mission, he was asking the crew to take as many photographs as they could of the earth—the 41-C crew had got very few pictures of the ground, and he suspected that the 41-D mission, which would be oriented toward the sun most of the time, wouldn't have the opportunity. Aside from that, he planned to leave the 41-G crew very much on its own about what to film in space. Judging from the film brought back by 41-C, astronauts do very well when they are left alone.

Integrated Sims

*T*here was about a month left before launch, which had been postponed from October 1 until October 5 because the 41-D mission had been delayed. The integrated sims for the 41-G crew had started in earnest about two weeks before. Browder told me that he directed most of the stand-alone training at these sims, for he regarded them as almost as challenging in their way as the flight itself, and in some ways more challenging. The Space Center was hopping; not only was the last phase of training for 41-G frantically under way, but the revised and revamped 41-D mission was finally in orbit. All around the Space Center, closed-circuit television was broadcasting occasional press conferences and continual air-to-ground commentary. The main object of 41-D was to unfold in space a solar array a hundred feet long, the progenitor of ones that might some day adorn a space station or certain power-hungry satellites. What captured the most attention, though, was the formation of an icicle, eighteen inches long and probably weighing about twenty-five pounds, at a water vent on one side of the orbiter. It was the source of two worries. First, it prevented elimination of waste water, including urine. (Indeed, there had been such trouble on past missions with the waste management system, which had virtually never worked properly and which had rendered the interior of the spacecraft so unpleasant, that Crippen announced that if he developed a similar ice problem with his seven-person crew, he was going to bring the *Challenger* back to earth at the next de-orbit opportunity. He was, apparently, at least halfway serious.) The other worry was that, during entry, the icicle might come loose and smash into a maneuvering-system engine pod. Sally Ride had once again been summoned to a simulator; she had spent most of the night of Sunday, September 2, there, seeing if she could reach the vent with the arm and knock the ice off. She could; the information, together with advice on how to do this, was radioed to the spacecraft. Henry W. Hartsfield, Jr., the commander, did the job successfully. Long after the icicle was gone, it lingered in people's imaginations. It would become the subject of sims. And Browder would write a memo to one of his superiors, suggesting that the arm in the future be equipped with a parabolic mirror so that it could focus the sun's rays on any bergs, vaporizing them.

Even though it was Labor Day, everyone connected with the 41-G mission was working. The astronauts had fallen farther behind. The sim crashes had continued. "We're at the point where we've lost so much time, we can't afford to lose any more," Ted told me when I saw him. "If we hadn't been so far ahead earlier, we'd be in serious trouble now." In the simulator sessions, he was now—as he had said he might —going directly to the most important lessons so that critical material wouldn't be left out. The crew, he said, was up on nominal training, which in the syllabus included ascents, entries, aborts, and some orbital material, which have predictable and repeatable flight plans. Where the crew was behind was in what he called "supplemental" problems—complex situations where systems malfs were piled one on top of another in a manner that was unlikely to happen, at least exactly that way, in flight, but that familiarized the astronauts with the orbiter's quirks and complexities so that when and if the unpredictable *did* happen, they would be in a better position to cope with it. About one-third of this material had not been covered.

Ted had fewer stand-alone sims with the crew, because the integrated sims now took most of the astronauts' time. These were behind schedule, too, adding to the pressure. The 41-G integrated sims were supposed to have begun the first week in August, after the arrival of the new, updated computer training loads. There was no point in doing the integrateds with loads designed for the *Columbia,* for its systems were sufficiently different in many respects from the *Challenger's* that the flight controllers, at whom the integrated training was principally aimed, and who had to deal with the orbiter's systems in great detail, could be misled. However, when the new loads finally arrived, they turned out not to be designed for the *Challenger,* either, but for the *Discovery,* the third orbiter in NASA's fleet, which was currently in space with the 41-D mission. Because the second set of loads was wrong, the integrated sims had had to be postponed until August 17; since that time, the astronauts had been doing as much as thirty-six hours of integrated training each week, working late into the evenings, over weekends, and on holidays.

There was, however, some good news: the crew in Browder's opinion, had finally crystallized, something that often didn't happen until after the integrateds had started, and that might have been abetted by the added pressure the crew and everyone else was under. The crew was now on stage in a way it never had been before, under the helpful but nonetheless critical eyes of a couple of dozen flight controllers in the Control Room, plus scores more in the various support rooms. Mistakes are not only pounced on, they can also hang up about a hundred people. Such a situation, when it becomes an everyday occurrence, cannot help but have a galvanizing effect on a crew. For whatever reasons, the crew had crystallized. "We started to see it during the standalones, but it really occurred during the integrateds, which are one more level of reality," Browder told me. "The astronauts are in harmony. They know each other's capabilities, what each one is supposed to do, and how they can back each other up or stay out of each other's way. The conversations are a lot crisper; when you bring in Mission Control, they have to be. And this is true not only of the conversations between the spacecraft and the ground, but among the crew inside the spacecraft. There is a lot less talk among them now. Crip will ask a question, Jon will say a couple of things, and they will both go and act. Crip no longer has to ask what procedures Jon has followed—there is complete confidence between them. And Jon, by the way, is *there*. He handles the maneuvering and thrusting systems as if they were second nature. He isn't fooled by sensor problems any more. He never misses these things now, however much we disguise them. He sees through our scenarios before they're completely played out! Crip spends less time monitoring Jon now; he treats him like a pilot. Which he is." Browder had once described Jon to me as the "measuring stick" for the mission: "As pilot, he naturally started the greenest and had the farthest to go." The measuring stick and what was being measured—the crew itself—had reached the top mark at the same time. Indeed, the crystallizing of the crew and the arrival of Jon at the top mark doubtless had everything to do with each other: one couldn't have happened without the other.

Crippen was not as concerned with the arrears in training as Ted was. Most crews are behind at this point; there is always a last-minute rush in training. Besides, Team Leads are always more concerned with filling in squares on a lesson chart than are commanders. "Crip knows what a crewman needs, irrespective of squares," Ted said. "We, with our squares, are not that far from Crippen's own estimates. We may feel that we should give them perhaps a quarter more exposure than Crip thinks they will need, but finally his own book is the envelope of knowledge that the crew will need to fly." Ted meant that Crippen, with all his experience, was the final arbiter of when the crew was ready. And Crippen felt that the crew was virtually ready to fly right now. "I don't think I'd get much sympathy at NASA headquarters if I asked for a delay in the launch date on the grounds that all the squares weren't filled in," Browder said.

*T*he crew's preflight press conference was held on Labor Day. It was the first time a preflight conference had been held when the previous crew was aloft, a sign that the pace of space flights was quickening. Crippen had told me it would be my last opportunity to see the crew personally before the mission, because of the extra pressure of their training schedule. (I would, of course, continue to attend sims.) It was the first time I had seen them in clothes other than blue jeans or corduroys. Robert Crippen and Jon McBride were wearing dark blue suits, and David Leestma was wearing a tan one; all three wore neckties—a sure sign that the astronauts were in the hands of NASA's Public Affairs Office. Sally Ride was wearing a stylish, all-white jump suit with a black belt—one would have expected one like it to be found in the wardrobe of Princess Leia in *Star Wars*. Kathy Sullivan, on the other hand, was wearing an old-fashioned, frilly white blouse and a white jacket that gave her a rather nineteenth-century look. The two payload specialists were there, too. Marc Garneau, the Canadian astronaut, was dark and square-cut, while Paul Scully-Power, the navy oceanographer, had sandy hair and beard and the continually jolly look of an old salt. With the exception of Sally, everyone on the crew had a naval, or at least a nautical, background.

The press conference went off in good form. They were, Crippen pointed out, the largest crew of astronauts in history—equal to the entire original astronaut corps of seven. It was, indeed, a crowd, Crippen admitted; but naval people were used to getting along in small quarters. There were questions, such as one addressed to Sally and Kathy asking how they felt about the fact that the first American woman to walk in space would be watching a man doing the work. Sally, and then Crip, explained that Kathy and Dave were partners working together, and that just as Dave had the lead with the orbital refueling system, Kathy had the lead for any emergency repairs that might come up.

Afterwards, I talked briefly with Garneau and Scully-Power. Scully-Power already knew the crew and most of the other astronauts because he had at various times lectured all the shuttle crews about what to look for in the oceans. He expected to spend most of his time at whatever window was vacant—among other things, he planned to keep his eyes open for what are called spiral eddies, which are fairly small (ten or fifteen miles in diameter, as opposed to larger, circular ones, which can be one hundred fifty miles across) and very numerous; nonetheless, they were not discovered until 1973, by the Skylab astronauts, and their dynamics have remained something of a mystery. He would also be looking for signs of solotons, which are waves that occur well below the surface, between two layers of water of different temperatures; they can be detected by telltale ripples on the surface. Scully-Power did not expect to have trouble adapting to the crowded conditions aboard the orbiter; he had served aboard submarines, which are crowded, too.

Garneau didn't expect to have trouble with the crowding, either, though he said he was having trouble fitting the equipment for the seven Canadian experiments he would be doing into the one drawer he had been allotted. They were a mixed bag. He would be doing a variety of medical experiments, and he would be looking for all sorts of glows—auroras, marine bioluminescence in the oceans, earth glow, and even a mysterious glow that occasionally appears on the orbiter's tail and is thought to be caused by ionization of oxygen atoms escaped from the earth. With his recent arrival and the problems of stowing his gear, he said he felt like the last man to join the crew of an ocean racer—not an enviable position, because the rest of the crew would already have gotten to know each other and possibly even taken the best bunks. (Garneau knows this because he has sailed in several races from North America to Bermuda.) Though they would certainly help out with chores like food preparation and cleaning up, there was some indication that Crippen was not prepared to regard them as full-fledged members of the crew. Graeme Ferguson, head of the IMAX company, had asked Crippen whether one or the other of the payload specialists might not be trained in using the IMAX camera; Crippen told him they were not to be used for that purpose.

The payload specialists were indeed a new breed of person in

space—more passengers than astronauts. For this reason, their training was limited to what NASA thought such people would need to
know in order to look after themselves and do their work in space. Two
months' training was thought to be sufficient; in that time, they would
take classes in habitability, they would fly aboard the KC-135 to
get some experience in being in weightlessness, and they would go
through emergency egress training with the crew at the Kennedy
Space Center, where there is a mock-up of the orbiter's crew compart-
ments, presumably water-tight, in a pond. Now the payload specialists
were spending most of their time with the crew in the simulator doing
integrated sims; even though they had no active roles, NASA thought
it was essential that they be familiar with everything that would happen aboard, if only to teach them when, and how, to keep out of the
crew's way. This would have another effect as well, which should perhaps have been foreseen: just as the simulations welded the crew together in the first place, they would succeed in integrating the two
payload specialists into it—not as full-fledged crew members, to be
sure, but as shipmates nonetheless. When the payload specialists first
went into the simulators, Browder felt that the crew behaved differently, initially at least. The five astronauts themselves, of course,
had so recently crystallized that assimilating two new people may
have been all the harder. "It wasn't that the two new members inhibited the crew, exactly; they were just as spontaneous," Browder told
me. "But Garneau and Scully-Power did add to the crowding. They
were two more voices to be heard on the communications loop. They
meant two more meals to fix. If there was a malfunction on one of Garneau's Canadian experiments, it was one more thing Crip had to bear
in mind. But after a while in the simulators, the crew began to accept
them. Now Marc and Paul go to all the crew's social functions. It's just
like adding two more people to the station wagon."

*I*ntegrated sims differ from stand-alones, of course, in that their purpose is to train the flight controllers as well as the crew and to inte-
grate the two groups. As far as operating the spacecraft is concerned,
the crew is already supposed to be trained. "That is, the crew is ready,
assuming it never has to talk to the ground," Mel Richmond, one of
four Simulation Supervisors, who for integrated sims are analogous to
Team Leads like Browder, told me once; he had conducted the integrated sim I had sat in on a couple of months before. Because the flight

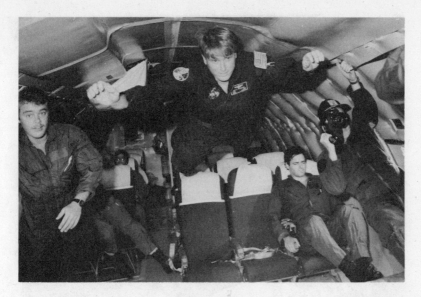

Jon McBride practices being in zero gravity aboard a KC-135 aircraft, which is flying in giant rollercoaster-like parabolas; each time the plane goes over the top, everyone aboard experiences about thirty seconds of weightlessness. As in a rollercoaster, they may also experience nausea.

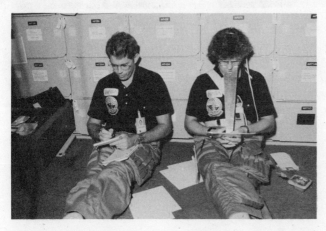

Dave and Kathy sitting on the middeck floor in the one-G trainer, an exact replica of the interior of the orbiter where astronauts practice such things as preparing meals, stowing equipment in cabinets, disposing of trash, operating experiments, and photography. Here, Kathy is practicing a medical experiment for testing vision in space, while Dave records the results.

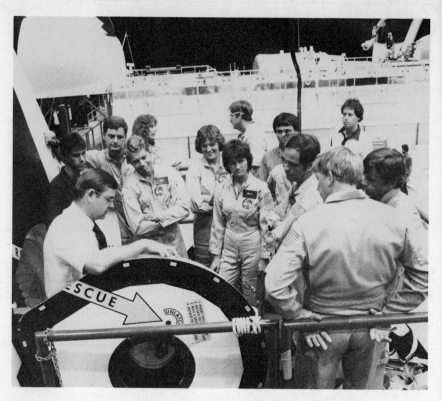

The 41-G crew stands outside the one-G (for one gravity) trainer, where they are about to have a training session for the launch and landing phases of the mission. Behind them is the manipulator development facility, for one-G simulations with the arm. Dave, Kathy, and Sally are facing the camera, while Jon stands at the right with his back to the rail. Marc Garneau, a payload specialist, is behind and to the left of Dave. Photograph taken for NASA by Otis Imboden.

controllers are continually flying all the missions, they in fact are pretty well trained already. However, there is a turnover rate of about 15 percent per year among the flight controllers, and as the old ones go, the new ones have to be trained. And the missions are different enough from each other in orbit (in particular, the payloads to be operated there) that the flight controllers have to learn the activities of each mission from scratch, in the manner of an experienced symphony orchestra learning a new piece of music. The astronauts, too, are at the same point; both groups from now on proceed together to master the mission, in greater detail than before. But the integrated sims are much more than training; they are a series of final rehearsals for the mission. At last, everyone most concerned with flying it—the astronauts, the flight controllers, and the experts for the payloads—is brought together for two months of intensive concentration on the mission specifics, during which time the score itself (the crew activity plan, together with various mission rules) will be submitted to the severest scrutiny and revised in light of any problems the integrated sims may turn up. In particular, the mission rules that John Cox, the flight director, has been working on will be tested. These are the hundreds of regulations governing such things as under what circumstances the satellite might be deployed, or the arm should be jettisoned, or the mission aborted.

There are about seven hundred flight controllers, with approximately one hundred sixty assigned to any particular mission, and these are divided into three shifts of fifty or so; for special problems, others can be pulled in. Of the fifty on any given shift, perhaps a dozen sit at consoles in the Mission Operations Control Room, where, over what is called the flight director's loop, they are in direct voice contact with one another and with the flight director and the capsule communicator, who is always an astronaut and the only person on the ground to talk directly to the crew. The sixteen-odd consoles are in a big, windowless room that resembles a movie theater and is suitably hushed; they face several screens at the front on which are displayed various types of data. The central screen, the biggest, normally displays a map of the world with the current orbit and the next two orbits marked on it; a replica of the shuttle moves slowly along it so that the controllers can tell exactly where it is in relation not only to the oceans and continents but also in relation to the relay satellite and the various ground stations, all of whose areas of radio contact with the spacecraft are marked out. The console positions represent a serviceable breakdown of the orbiter and the mission, in more detail than the four instructors' consoles in the mission simulators. The consoles in the front row, called the trench, have to do with aspects of motion: the flight dynamics officer is concerned with ascents and descents; the trajectory officer is concerned with orbits, refining them, or changing them while the craft is in space; the guidance officer and the guidance, navigation, and control officer are concerned with navigation and computer programs for that purpose; and the propulsion officer is concerned with the state of the rockets and their fuel. In the second row, the data processing system officer is concerned with the computers; the payloads officer is concerned with those items; and the electrical and environmental control manager is in charge of those two systems. At the left end of the third row is the officer in charge of radio communications, both voice and telemetry; next is the flight director; the capsule communicator; and, last, the flight activity officer, who is in charge of the crew activity plan and any changes in that massive document that may come up during the flight. Most of these console positions are backed up by the other thirty-odd flight controllers on the same shift, who sit in a number of support rooms nearby; most of the console positions have their own support room, so that if the electrical and environmental control manager has a problem with the electrical system, or if the data processing system officer is stumped by a computer glitch, he or she can get their associates in their back rooms to come up with the answer.

Behind a large glass window at the right front of the Mission Operations Control Room is a smaller room called the Simulation Control Area; it is here that the Sim Sup and the other instructors running the integrated sims sit. They have their own consoles and although these are laid out differently from the ones in the main room and are closer together, the jobs of the men and women at them roughly parallel the positions of the flight controllers—that is to say, each flight control position has its own gremlin in the control area. The team of instructors for the integrated sims consequently is much bigger than for the stand-alones. The chief gremlin—the Sim Sup—is, of course, the counterpart of the flight director, and also of the Team Lead over in the mission simulators. The stand-alone instructors—Browder and company—are at their own consoles to help the integrated team by carrying out its instructions with their light pens, as I had seen them do before; they are the only ones who can operate the simulator computers. Connecting all these people are a number of separate communications loops: the astronauts' loop, of course, over in the simulators, along with the stand-alone team's; the flight director's loop, presided over by that individual; and loops for each of the console positions, connecting them with their own back rooms. And then, of course, the instructors in the control area have their own loop, presided over by the Sim Sup. The Sim Sup and his associates can punch up any of the loops they want. Usually, as they eavesdrop on the astronauts, or the flight controllers, or the back rooms, they have smug little smiles on their faces at the trouble they are causing. The window between the Simulation Control Area and the Mission Operations Control Room has a curtain that is open during the sims, but is apt to be tightly drawn during the actual mission, as if to obliterate any memories the flight controllers might have of unsolved malfs, anomalies, glitches, nits, or bites. (There have been times during sims when a flight director has had the curtain closed, to blot out a particularly dirty trick.) The people who work in front of the curtain—the flight controllers—think of themselves as living in the real world, and they talk of the area behind the curtain as the "sim world," as if it were a land of dreams—bad dreams. With all their projections of what might go wrong, much of it derived from what actually *has* gone wrong on previous missions, the denizens of the sim world represent a sort of collective unconscious of space flight, and a very mischievous one at that.

*T*he Simulation Supervisor for the 41-G mission was Edwin Rainey, a trim, soft-spoken man in his late forties who used to be a pilot for NASA at Ellington and then became a Team Lead for stand-alones before moving to integrateds. (People at NASA tend to stay a long time, often moving about within NASA's ample infrastructure.) Rainey invited me to attend a script-writing session with his team— one of four teams of integrated instructors—something that Browder had never felt free to do because the stand-alone teams are much smaller, and the flow of their creative juices might be blocked by an intruder. Rainey's team, on the other hand, had about a dozen members, a large enough group to generate an unstoppable flood of creativity.

I met Rainey in his office on the second floor of Building 4, where he and the three other Sim Sups had desks; it was a few doors down from where Browder had a desk in a more crowded office. Rainey and Browder knew each other—they passed each other in the hallways and met at the vending machines downstairs—but they had not consulted much on the 41-G crew. Browder had not briefed Rainey on any special strengths and weaknesses of the crew. "It's not really a handing over; Ted is still very much involved," Rainey said to me. "He did not tell me any significant areas we need to work on, nor did I ask him." This was by no means an oversight on Browder's or Rainey's part, for from long experience both men know the Team Leads have been working with a crew for so long that they might be blind to their weaknesses, or if not, that they might not want to reveal them because the Team Leads might not want the Sim Sups to trip up the crew—*their* crew—and make them look bad. Crippen, too, had made only minimal inputs: Rainey had met with him and John Cox, the flight director, three months before, very briefly, to discuss the general ground to be covered by the integrated sims. Cox, however, had considerably more input, since the integrated sims would be aimed particularly at the flight controllers. At the meeting, Cox had handed Rainey two or three pages of mission rules which he wanted pushed during the sim. A good deal of analysis over several weeks had gone into making the rules, and a main purpose of the sims was to see how they worked out in practice; many of the rules could conflict with one another and in various ways needed ironing out. Periodically Rainey checked with Cox, for it was in the interests of both men that the list be thoroughly covered. Rainey had considerable latitude in how he worked his way down the list—he used it largely as he saw fit, and he added several topics of his own. His other source of inspiration, of course, was the crew activity plan, which the integrated sims were also meant to test.

The integrated sims would largely be a fresh start for everyone concerned, as the mission itself would be. The scripts are much more detailed than the ones used in the stand-alones, because the flight controllers receive much more data from the spacecraft systems, via the communications data link called telemetry, than is available to the astronauts. The greater amount of data, of course, explained why there was a greater number of flight controllers as well as integrated instructors—and also why the integrated sessions last so much longer than the stand-alone sessions. Later, Keith Todd, the manipulator arm instructor (who took part in the 41-G stand-alones as well as the 41-G integrateds), told me, "In the integrateds, there's a whole other level of detail. In the stand-alones, with just the crew, I only have to make them react to the knowledge they have—and to react fast. But in the integrateds, we're challenging the flight controllers, with their additional data and their access to the orbiter's blueprints. In the stand-alones, we make the crew work through the checklists, procedures, and manuals they have aboard, but in the integrateds we shoot for cases that will drive the flight controllers right straight back to the engineering drawings." The scripts, he said, were not only more detailed, but there was more interrelation between the different console positions. "Here, we have to play together more," he said. "In the stand-alones, if I want the arm to fail, I just fail it myself; but in the integrateds, I might get someone else to fail an electrical bus, so that the arm will go down." Consequently, when the scripts are carried out, they are followed more closely: because of the greater number of people involved and the greater dovetailing of their interests, one instructor's bright intuitive flash during a sim is more apt to upset another instructor's well-laid plans. There is still room for improvisation, but it has to be done more carefully.

The script that was being prepared was for 41-G's three-day sim—it would last fifty-four hours, with another four or so for the debriefing afterwards. Because of the duration of three-day sims, there is always a problem fitting them into the schedule, particularly so later in a crew's training, and there were scheduling problems now. Technicians at the Cape were a week late replacing some of the *Challenger*'s insulation tiles, and as a consequence a three-day rehearsal of the countdown before lift-off—the terminal countdown and demonstration test —scheduled for the next week had been bumped into the following week. That meant that the long sim most likely would be moved a week earlier, which would make it *next* week. This could mean that Rainey and his team would have to work over the weekend. "Schedules around here are cast in Jell-O," Rainey said.

He led the way from his office to the big conference room just down the hall on the second floor of Building 4. About a dozen instructors, mostly in their twenties or early thirties, were already there, seated around a square composed of several big wooden tables. Rainey, who had the breezy, laid-back air of a wing commander in a World War II movie addressing his pilots about their next mission, stood in front of a blackboard and explained laconically about the change in schedule, about the Jell-O, and about the possibility of working over the weekend. There were groans. Then he talked about the sim: it would begin in the second day of the mission and run into the fourth day. The background was that the astronauts would have failed to launch the radiation budget satellite as scheduled on the first day, which would mean that according to the crew activity plan the first order of business (before breakfast on Day 2) would be to try once more to get it off. After that, if they were successful, the astronauts would fire their maneuvering system engines and descend from their orbit of one hundred ninety-eight miles, the optimum for deploying the satellite, to one hundred forty-eight miles, where the shuttle-imaging radar antenna would be deployed. Rainey and his team would be testing some of the concerns I had heard expressed at the flight-readiness review I had attended in February. He handed out to everyone copies of the timeline for those days, which had the major known events marked on it, such as sleep periods, meals, and when the spacecraft would be in sunlight and when in night. The timeline otherwise was relatively empty looking, resembling blank sheets of music, and it would be the purpose of the meeting to fill in the timeline with a vast number of nits, malfs, anomalies, and glitches, in quantities resembling a scherzo passage in a Gilbert and Sullivan operetta. There would be harmonies and dissonances, with one theme leading into another. There would be the tinkling of minor nits and the crashing of major malfs.

The first order of business—as is the case with writing any script—was to establish the setting. This meant establishing things that happened on the previous day, Day 1, before the curtain went up, and why it was that the radiation budget satellite had not gotten off then. Ideas came thick and fast, the way they are supposed to do when a musical comedy is being written by a team of quick-witted playwrights and songsmiths.

"Maybe they lost the relay satellite for some orbits. They'd never deploy without that because the ground crew needs it for checking out the radiation budget satellite."

"Why was the relay satellite down?"

"Its batteries went dead."

"But we'll need the batteries and the relay satellite later."

Silence.

"Maybe the brace release on the shoulder joint of the arm failed, so the arm couldn't be unberthed."

"Maybe there was a big dust storm at White Sands which required folding up the big antennas." White Sands, New Mexico, was the ground station for the relay satellite.

Everyone liked that idea, so Rainey jotted it down on the blackboard. He also jotted down the failed brace release that secured the shoulder joint for launch. In the sim world, there might easily be more than one explanation for the failure to launch the satellite, and in any event it started a new subplot, a new minor theme.

The meeting broke up into small groups, and Rainey moved busily from one group to another, occasionally writing something else on the board—"inertial-measurement unit down"; "auxiliary power unit gearbox failure—sensor problem." Malfs were quietly being pulled out of the air all around the room. The instructors were flowing about, conceiving a situation here, checking to see if it interfered with somebody else's subplot there.

A blond young man, Stephen Hamm, who was in charge of the onboard computers (in charge, that is, of fouling them up) said to Rainey that he wanted to take out two inertial-measurement units later in the morning of Day 2.

"They'll be fairly busy until after Day 3, when they'll finally get the satellite launched. So why don't we do that then?" Rainey suggested.

"That's great!" Hamm said. "That means they'll lose the measurement units at the same time as they're trying again to launch the radiation budget satellite." The creative juices were clearly running high. I asked Rainey to what extent he and the others thought of themselves as collectively writing a play. He said, "I don't know what playwrights do, but I *do* know that we get very caught up in our scripts."

"On Day 3, I want a White Sands computer crash, as well as a measurement-unit failure," Hamm said to the room at large. "Is that O.K. with everyone?" It was.

Ron Weitenhagen, an engineer in his mid-thirties with a high forehead and a Lincolnesque beard, who was in charge of throwing monkey wrenches into the workings of the OSTA package and the imaging radar antenna, was especially pleased. "With White Sands and the relay satellite down again, they'll have to record the radar data onboard," he said. He began thinking up ways to screw up the tape recorder. There was clearly no limit to the way problems could feed on each other.

Problems fed on each other, not only because doing so could happen in the real world, but also because one of the ulterior motives of the sims was to get many different people, in different back rooms, talking to one another, to open channels of communication among them that might be useful during a real emergency. Rainey was talking to a short, stocky engineer, Jack James, who was responsible for making things break in the propulsion system; James was trying to persuade the communications instructor to put some misinformation about the maneuvering system rockets in the telemetry data so that when the astronauts, following the successful deployment of the radiation budget satellite, were trying to get down to the lower orbit, there would be confusion in many quarters. "The propulsion flight controller will beat on the communications flight controller to make sure that the data is correct," James told me, a dreamy, far-away look in his eye, as though he were a dramatist trying to rough out a complex plot. "And the imaging radar guy will beat on the flight director to get him to circularize the lower orbit as fast as possible—right now, they will be in an elliptical orbit as a result of the first burn. [Generally speaking, the spacecraft stayed in orbits as nearly circular as possible to keep the radar and large-format camera images in focus; to get from one circular orbit to another, the maneuvering system engines would fire once, putting the spacecraft for a time into an elliptical orbit; then, as we have seen, it would fire a second time, to circularize the orbit at the new altitude.] That means that the trajectory group will have to keep track of the orbit, and keep recalculating the burn they need for circularizing. And the comm people will have to keep recalculating the ground coverage—the ground stations it will be passing over when it is out of touch with the relay satellite. So I hope this one problem will make a lot of people jump and get them arguing with each other." James clearly had as great an ability to project a scene as anyone in the room.

A group of three or four instructors, including Keith Todd, were putting their heads together to figure out exactly why the radiation budget satellite would not get off on Day 2. The dust storm at White Sands that had prevented the deployment on Day 1 had been allowed to subside—in the sim world the same anomaly cannot be used twice, lest it appear that the creative juices had stopped. A young engineer with curly black hair, Sonny Garza, who was in charge of making trouble for the radiation budget satellite, was telling an older engineer, Carl Keppler, with short silvery hair, who was in charge of glitching the electrical system, that he had just decided that the satellite deployment on the second try would be put off once more because its batteries could not be charged. The batteries had to be charged just before release (if it were done earlier, they would lose their power in the cold of the cargo bay), and this was done by an umbilical from the orbiter that provided electricity and communications with the satellite while it was in the bay. (Once it was on the arm and out of the bay, it would sprout a pair of solar arrays which would keep the batteries charged from then on.) Garza was seeking Keppler's collaboration; he wanted him to cause a glitch in the orbiter's electrical system so that the re-charger wouldn't work.

Keppler was interested. He said, "Tell you what we'll do: we'll cause one of its circuit breakers to pop. To do that, we have to have something else shorted back here, upstream." He pulled out a chart of the electrical system, and the two of them pored over it. Evidently, the sims forced not only the flight controllers to the drawings, but the in-structors as well; the deeper level of detail at times baffled everyone.

An associate of Keppler's, Mark Suarez, located a likely piece of equipment to short out. "You decide what you want us to do," he told Garza. "Do you want them to be able to reset the circuit breaker, or not?" (Almost always the pronoun *them* referred to the flight con-trollers who had to solve the problem, not to the astronauts, who would not have the detailed information.)

"Let's have them reset it, and it'll go bang," Garza said with a laugh. (Even though *them* now meant the astronauts, the astronauts would be carrying out the flight controllers' instructions and were their surrogates.)

"O.K.," said Suarez. "It'll be a slow burnout, and they won't be able to use that piece of equipment again. So they won't be able to re-set the circuit breakers. So they can't recharge the batteries. Is that going to stop them from deploying the satellite?" I felt as though I was watching the booby-trapping of a long, sinuous trail.

Robert Bassham, a young engineer who, with his black hair and deep tan, looked remarkably like Keith Todd, and who worked with Garza (whom he also somewhat resembled) on the radiation budget satellite, said, "No, it won't. They might decide to chuck it overboard quickly, while there's still some power in its batteries. But we can fix that with an umbilical failure. It'll be stuck—they won't be able to detach it from the satellite! We'll have to figure out why."

"That's a great idea," said Garza. "It'll force them to think about going EVA to unplug the umbilical, or waiting until Day 5, when they're going EVA anyway. And if they wait until Day 5 before deploy-ing and going down to the lower orbit, the imaging radar people, who want the lower altitude, will be after them with knives."

"If they keep the satellite in the bay that long, they'll have to con-sider that its hydrazine might freeze and leak. It's not a good thing to keep in the bay—another reason they might decide to get rid of it right away," Bassham said.

"And they wouldn't have to go EVA to do that," Keith Todd, the arm man, said. "They could use the arm to pull the umbilical." (This was an idea that had recently been developed in the simulators. The umbilical is a sort of extension cord for electricity and telemetry that plugs into the satellite from the orbiter. If the umbilical failed to unplug automatically, the arm could either bump the umbilical to jar it loose or, failing that, grapple the satellite and just pull—it would be like unplugging a lamp by picking it up and tugging the whole thing.) Todd thought he might be able to put a glitch into the arm's end effector to keep them from doing that.

I talked with Todd about some other plans he had for the arm. He was going to break its automatic mode so that it would work only in the single-joint mode, in which the astronauts controlling it would make each joint move individually instead of in unison, and hence certain maneuvers that depended on the automatic mode, such as inspecting the water vent with the wrist camera to see if ice had formed there, would be more difficult. Almost any new anomaly that occurred in space, such as the 41-D icicle, had an inspirational value in the sim world.

I asked Todd what prompted him and the others to pick some of the anomalies they had chosen. "Different reasons, none in particular," he said. "I want to break the single-joint mode because I haven't done that for a long time. The last time I did it, I got good work out of it; it made the flight controllers rethink their procedures for running the arm." In integrated sims as in the stand-alones, things often happened for no other reason than that the instructors wanted them to.

Todd stopped because he heard Suarez saying to Keppler that he planned to kill a couple of the spacecraft's main busses. Both had moved on to greener glitches.

"What kind of busses are you killing?" Todd asked.

Suarez told him.

"You're sure it'll have no effect on me?" Todd asked suspiciously. He wanted to make sure he would still get power in the arm.

"Not as far as I can tell," Suarez told him.

Todd told me, "In a script-writing session you can't be too careful—somebody might screw you up without meaning to." Clearly, in the sim world it was O.K. to screw up an astronaut or a flight controller, but not a fellow gremlin.

*T*he next day, I joined Rainey and his team in the Simulation Control Area, behind the glass window at one side of the Mission Operations Control Room, to watch an eight-hour integrated simulation of Day 1. The three-day sim had been restored to its original time two weeks later, and everyone seemed relieved. In the control room, the flight controllers were sitting quietly at their desks, or in some cases standing and talking quietly to each other, tethered to their consoles by their headsets. Though the control room with the big screens in front looked like a movie house, watching it was not very good theater; the only action was the slow movement of a pink silhouette of the spacecraft along a pink line representing its orbit, across a map of the world projected on the biggest screen. Even in a crisis there is nothing more to see; all the action takes place over the communications loops, and even that can be terse and difficult for an outsider to understand. There was much more action in the Simulation Control Area, where Rainey's people were drinking coffee and talking to each other without the benefit of headsets. Sometimes they put on their headsets and talked to Ted Browder, Shannon, Andy, or Randy across the duck pond in Building 5, who were also drinking coffee. On the loops, Rainey and Browder addressed each other as "Team Lead" and "Sim Sup." The relationship of the two men, and the two teams, was complex. On the one hand, the stand-alone team has to carry out the instructions of the integrated team. If the computer gremlin in the simulation area wants to pull down a multiplexer-demultiplexer, he calls Shannon and tells her to do it. However tempting it may be for Shannon and the other members of the stand-alone team to improvise malfs, anomalies, glitches, or nits of their own, they never do. "That would never work, because then Sim Sup would be in the dark, just like the flight controllers and the crew," Browder told me. There is honor among gremlins. Browder and his team acted as colleagues and consultants to Rainey and his team; their superior knowledge of the crew and what it was doing at the moment in the crew station was indispensable. They often made suggestions, which Rainey would take about half the time.

Nothing much was going on at the moment. The orbiter, I could see, was flying over the United States, well within radio coverage by the relay satellite, and Sally Ride was awaiting instructions to pick up the radiation budget satellite with the arm. Garza, one of the radiation-budget-satellite instructors, sitting to Rainey's right, had told his counterpart in the simulators to fail the talk-backs, or the sensors, which would tell the controllers on the ground whether the satellite was powered up—it was sitting in the cargo bay, supposedly drawing electricity from the orbiter through the umbilical. (Garza's counterpart in the simulators was Bassham; as had been the case in the stand-alone training, instructors with special knowledge of such things as the payloads or the arm were needed to do the light-penning.) When the flight controllers found they couldn't tell whether the satellite was powered up, there was consternation on the loop, in the form of lengthy discussions. Integrated sims, I had found, proceeded at a much slower, statelier pace than the stand-alones, approaching almost exactly the pace of a real mission. I tuned in on the astronauts, who apparently were also having their refreshments; they would let the ground worry about the arm problem.

Crippen was saying, "Anyone like a drink?"

"I could go for something in the lemon line," Sally said.

"Orange-mango," said Kathy.

Before going down the hatch to the middeck, where the food lockers are, Crippen asked Paul Scully-Power if he would like anything. There was no answer.

"Scully-Power, are you with us?" Crippen asked.

"Scully-Power is on his own personal comm link," Sally said. He was clearly undergoing a little hazing—perhaps he had dozed off.

When the crew had settled down with their drinks, the ground was still working on the nit. For the crew, the integrated sims were very different from the stand-alones. Rainey told me the crew had had no trouble making the transition; although in the stand-alones they had gotten very good at solving problems themselves, they had quickly gotten used to the idea of sitting back and waiting for recommendations from Mission Control. Among other things, it gave them more time to test out the space drinks. Rainey said the transition varied from crew to crew: some started off trying to solve everything themselves, as they were accustomed to doing, but others had the opposite trouble, waiting too long for Mission Control to come up with the answer and not taking enough initiative. Clearly, a crew had to learn the proper balance between being independent of, and dependent on, Mission Control. At first, a new crew might not be aware of exactly what Mission Control could do for them. "During stand-alones, someone might say, 'Now we'll talk this over with the ground'—only, of course, there isn't any. Here, they learn that the ground isn't something they should compete with, or be embarrassed if they make a mistake in front of. The flight controllers are their friends in court," Rainey said. One of the things a new crew had the toughest time getting used to in integrated sims, Rainey told me, were the periods when they would be in touch or out of touch with the ground—periods called loss-of-signal or acquisition-of-signal. Sometimes a new crew will be very restrained when it is in touch, working amicably with the ground; but then, when it is out of touch, the crew will rush in and solve problems as if they were back in stand-alone training. When they are in touch again, the ground might not approve everything that had been done behind its back. All these kinks normally get ironed out after the first couple of integrated sessions. The 41-G crew, though, from the start tended to fit right in with Mission Control, because Crippen had flown so many times before that he had long since established his relationship with the ground. Browder had told me once that he thought the stand-alones were as realistic as the integrateds, particularly in the case of time-critical malfunctions, because the astronauts were usually quicker at solving problems than the ground, and were usually ahead of it—after all, on a real space flight, the astronauts had the advantage of being there. Crippen disagreed. "In flight, you talk almost everything over with the ground, and talking to the ground turns out to be the thing that is real," he said. "In flight, unless something is time-critical, I always feel more comfortable talking it over with the ground, on the theory that the more heads you have solving a problem, the better off you are."

Rainey tuned in on the flight controllers' loop again. The problem had become more complex now, and more voices were chiming in. The umbilical itself had failed so that the radiation budget satellite, still in the bay, was getting no power from it. In its inert state it would be unable to detach itself from the satellite when the time came to pick it up with the arm, which was very soon. All this sounded similar to the satellite-release problem Todd and Bassham had been cooking up for the long sim: many scripts, I found, rang changes on the same set of malfunctions. About this time, there was a problem with the arm, too: the backup system for releasing the satellite into space—a sort of spring mechanism—had failed. Although the primary system for release was in good shape, there was a mission rule that prohibited picking up the satellite without the backup, for if the arm picked it up and couldn't let it go, either an astronaut would have to go outside and disconnect the end effector, or else the arm with the satellite on it would have to be jettisoned (there are explosive bolts for this purpose that could be fired from the rear control station). Rainey, Todd, and Garza listened on the various loops to what everybody was saying to one another. As they frequently did, they had tight little smiles. Rainey told me that he listened on the loops to make sure the right people were talking to one another; if they weren't, he took them to task in the debriefings afterwards. He said that sometimes, listening on the loops, where he could hear the flight controllers but they couldn't hear him, made him feel like God—as if he could peer into other people's minds without their being aware of it.

Anyone listening in for the first time on the communications loops, whether in a sim or during an actual mission, might easily think that he or she had somehow tuned in on the inner thoughts of one mind—of a single individual who is ruminating about having fallen into difficulties and what might be done to climb out of them, with different parts of the brain raising different questions or finding different solutions. Arm people were talking to satellite people, and satellite people were talking to orbiter electrical-systems people, and they were all talking to the flight director, John Cox. They were discussing getting the astronauts to recycle a switch for backup power to the umbilical and the satellite. They were discussing using the arm to unplug the umbilical—using the "pull-the-whole-lamp" method—in case the backup switch didn't help that problem. Most of the flight controllers seemed to be against doing that. "That's a very conservative position," Rainey said. Flight controllers tend to be conservative early in training for a mission until they develop more of a sense of what can be done with the hardware: the payload hardware in most cases would be new even to veterans. They decided that, with the backup release switch on the arm broken, they had probably better not try to deploy the satellite with the arm at all. Rainey felt that that was even more conservative. Yet both Cox and Crippen concurred. Ron Weitenhagen, the shuttle-imaging radar instructor, whom I was sitting next to, said to me, "The crew is being very quiet—it's going along with all this conservatism. When flight rules are concerned, the crew usually defers to the ground. The managers, responsible for the equipment, are here. If the astronauts go ahead and pick up the satellite—which they may well want to do—and they can't deploy it, and it's stuck to the arm, then they'd be to blame. They're the ones who'd have to go EVA to jettison it. So it's natural they'd be cautious." The flight controllers, of course, were following the same reasoning.

Over his headset, Weitenhagen called Bassham at the mission

simulators, where he was handling all the payloads, and asked him to fail a microswitch in the radar antenna. "If they end up not doing the satellite, they could deploy the antenna for the radar antenna right away, and I want that glitch in," he said.

Ed Rainey said to Garza, "Would you be happy if we got power back to the umbilical and fixed the arm so that they could deploy today?"

Garza said he didn't care.

Rainey took the question up with John Cox. Cox was the sort of big person who didn't move much but nonetheless seemed to be bursting with latent energy primed for action, should it be needed—many flight directors give this impression. Even though he himself was one of the flight controllers, and hence was being trained along with the rest of them, he was a veteran of many flights and his role in training was like Crippen's, in that he was constantly being consulted on the training and could take an active hand in the direction a sim might take. He perhaps exerted an even greater influence on events in the middle of a sim, for he could talk privately with the Sim Sup, while Crippen, in the crowded crew quarters, was not in a position to hatch plots with Rainey or Browder without a lot of people hearing him. Whereas Crippen had the final word on anything affecting the spacecraft, Cox had the final word on mission rules.

"There's no way we can deploy the satellite, the way things are stacking up," Cox said to Rainey.

Rainey asked him if he would be glad if they gave everything back and allowed the deployment.

"I don't care if they deploy or not," Cox said, giving the same answer as Garza. "It's a good problem! I hope they learn something from it—that they don't just sit there and cuss."

They agreed among themselves to restore the umbilical to health and allow it to finish recharging the satellite's batteries and also to detach; but they decided not to restore the health of the arm. This immediately meant that the flight controllers had to consider another option for deployment, and this provoked a lengthy discussion on the loop: with the satellite now powered up and apparently releasable, some of the orbiter propulsion people suggested a technique that previously had been analyzed and approved, that the satellite be released directly from the bay without the arm. The latches holding it down would be opened, and the orbiter would then simply drop away from it. This would be done by firing the orbiter's forward thrusters, thus slowing the big spacecraft so that it dropped into a lower orbit, allowing the satellite to float free, as though it were an orange crate on the deck of a submerging submarine.

As soon as Jack James, Rainey's propulsion man, heard this idea being discussed in the propulsion flight controllers' back room, he called Andy Foster, his counterpart over at the simulators, and asked him to put a sensor failure in one of the forward jets. "That way, if they decide to drop down from the satellite, the crew can't tell whether the jet is firing, though the ground can. It will provoke a little discussion between them." Writing a script, or ad-libbing in a sim, was in many respects like a conversation, with one person after another kicking the ball a little farther along, each from his own perspective.

However, the satellite controllers, who were at the Goddard Space Flight Center in Maryland, in the Remote Payload Operations Control Center, objected strenuously to the notion of the orbiter just dropping away from the satellite; if anything went wrong with it, such as a failure of its solar arrays to deploy, the satellite would be impossible to retrieve. The flight controllers in Houston, though, wanted to deploy by dropping away—the method had never been tried before. In the end, they decided not to do it. Rainey told me that he didn't blame the payloads people for being so conservative, because they built the hardware and are the ones finally responsible for what happens to it; at the same time, they are the least experienced and have the least confidence in the capacities of the orbiter. The discussion continued in the debriefing that followed the sim. Cox told the satellite representative at Johnson, who sat at the console directly in front of his own, that if the satellite people did not want to deploy by dropping away, then the mission rule allowing it as an option should be removed. However, the satellite people wanted to keep the rule, in the event of an emergency—say, that the arm wouldn't work and the satellite developed a hydrazine leak so that it couldn't be brought back to earth. Ironing out the mission rules, with everybody present, was an argumentative process.

Another satellite person said heatedly to Cox, "If there is an arm failure, we'd want you, Flight, to keep on trying to fix the arm—we can always delay the deployment." Such a delay, of course, would cut into the time for radar observations, but Cox went along with the satellite people's request (or rather, demand) because he felt getting the satellite off was more important than gathering a little more radar data. He also knew that, in a real mission, there was always the possibility of staying up an extra day to allow more time for the radar observations—a possibility that was of little consolation to the radar people, for in anything as tricky as space flight, a bird in the hand is always worth two in the bush. Without meaning to, Cox had opened a whole new can of worms, for the radar scientists, sitting in a control room a few yards away in the Mission Operations Control Center, objected strenuously to the low priority Cox had given their data. The scientists who have worked on instruments going into space get more wrapped up in simulations than even the regular flight controllers— they have not spent much time in NASA sims, and they are working with their own instruments and data. The operations director of the radar control room, William Brizzolara, said later that feelings in the payloads control center were often tense. "It's very real," he said, "and tempers flare. A scientist will fight for his hardware, or his time to take data, as much as for a real flight. Every time the satellite deployment is delayed, it's mayhem back there, as no one wants to lose his own data takes with the radar. Sometimes I've got to throw my hands in the air and say, 'Wait a minute, guys, it's only a sim.' " The lack of windows encouraged the illusion. Of course, there was a strong element of reality in the sims, for practices and priorities developed now would apply to the flight itself.

The radar scientists had two advantages over the satellite scientists: first, their control room was a stone's throw from Mission Operations Control Room; second, there were many more of them. Forty-four scientific teams were connected with the radar, and all the teams had at least one representative in Houston; they were using the radar to acquire data in such varied subjects as oceanography, forestry, cartography, hydrology, geology and archaeology. Kathy Sullivan was a member—she was studying the radar "signatures" of impact craters, in the hopes that others could be identified. Having her on the team gave it added clout. And a number of archaeologists and paleontologists were using the antenna to find medieval settlements on a Baltic island and evidence of early humankind in Kenya. (An earlier version of the present radar antenna, which had flown on the second shuttle mission in November 1982, had demonstrated the ability to "see" several meters underground in dry areas—it had revealed ancient stream beds underlying the Sahara, and subsequent investigation in those areas later unearthed evidence of ancient human habitation.) Because of the great public interest in all this, the imaging radar scientists had extra leverage at NASA. The pressure they exerted took many forms. For example, on that day a door opened at the left front of the Control Room—the door nearest the radar control center—and a scientist walked in and handed Cox a light blue T-shirt that had SHUTTLE IMAGING RADAR TEAM MEMBER written on it. Cox accepted it, though he didn't put it on. "Now remember, no more bad calls," the scientist said.

Clearly, it was high time to give the arm back, deploy the satellite as expeditiously as possible, and start gathering data with the radar. Cox conferred with Rainey. Rainey conferred with Todd. They agreed that the mission rules and priorities had already been well explored by the problem—it is, in fact, the conversations and the interactions between people that the sims are meant to foster, not the end result, which is often anticlimactic. The malf had served its purpose. They gave the arm back. The astronauts would not be doing the drop-away deployment. Jack James, who had put the glitch in one of the forward thrusters that would have been used for that purpose, was annoyed. Because at the moment the astronauts were using the vernier system (the smallest jets) for attitude control, James decided to foul up that system. "That's another way to force them to use the regular forward thrusters," he told me, placing a call to Andy Foster on the loop. In a sim, if people didn't fall into a trap, it was O.K. to shove them into it by other means.

With the arm, Sally Ride picked up the satellite and held it over the payload bay, where she could see it through the overhead window; the arm's elbow was bent and the wrist tilted, almost as though a human arm was poised to flick a paper plane across a room. While it was there, the ground continued to monitor it through a connection in the arm. At a word from the ground, Sally pressed a switch to deploy the satellite's two solar arrays. Nothing happened. "You probably noticed we tried to deploy the arrays and got no joy," Sally reported. ("Getting no joy" was one of the astronauts' favorite expressions; it means that something didn't work.) "When she tries it again, one array will pop up, but the other one won't," Garza told me. "That will make the ground consider whether the spacecraft can survive on one array or not. It's a question we posed once before and they didn't know the answer, so we're forcing the issue. At the same time, we're breaking one of the four latches that would hold the satellite down in the cargo bay, if they decide not to try to deploy but to bring it home. There's a mission rule that says you can't bring it back with only three latches. It's another testing of the mission rules. They'll have to decide what to do, and I don't know the correct answer."

To Garza's surprise, both arrays popped into place. Bassham, over in the simulators, had overruled Garza and not put the problem in— something that Browder and his team would not have felt they could do. It was getting late, and he wanted the astronauts to deploy the satellite before the end of the sim. Sally deployed it perfectly, with almost no tip-off rates. Rainey felt that Bassham had been right to allow the release—it was, in a way, a reward at the end of a long afternoon. Although such rewards do not always happen in the real world (or, for that matter, in the sim world), integrated sims are so realistic to everyone involved that astronauts and flight controllers, just like controllers for the payloads, can be genuinely disappointed when matters don't work out. During the three-day sim of the STS-7 mission, when all the technicians for the SPAS satellite from Germany had been present, Sally had been unable to grapple the SPAS with the arm (the error was not hers but the software's), and it was lost. "At the beer bust afterwards, they were so sad!" Browder told me. "Yet the real SPAS was safely at the Cape." Astronauts who have made serious errors in sims, possibly endangering a mission, have come down from the crew stations in the simulators looking quite shaken. As today's sim had gone well, Bassham did not want to distress anyone. The line between the real world and the sim world is very thin.

*L*ater that week—the day after the 41-D mission returned to earth— Dave and Kathy had another session in the weightless environment training facility. They arrived at the edge of the big pool with broad grins on their faces and announced, "We're next." They were in a good mood. Nothing could faze them—not even the news that the previous month a Russian woman, Svetlana Savitskaya, had beaten Kathy as the first woman to go on a space walk. Kathy told me she couldn't care less. "When I was first assigned to the EVA last November, I said at the time that the Russians would probably try to do it first, and it doesn't surprise me that they have," she said. "And it doesn't bother me now in the least. I haven't met Svetlana. I hear she's an impressive person, with an impressive flying record. It would be neat to meet her. Maybe she'll still be in the Soviet space station when I'm up there." Svetlana Savitskaya, who had been in space once before, had also beaten Sally to be the first woman to go into space twice, and when I asked Sally how she felt about that, she laughed and said in genuine wonder, "I never thought of that being a record; it honestly never occurred to me. It's clearly a major milestone!"

Dave and Kathy had recently overcome a more serious threat to their EVA. Normally, before an EVA, astronauts do two or three hours of what is called "prebreathing": because the cabin atmosphere is 80 percent nitrogen like the earth's, they have to breathe pure oxygen in order to get rid of the nitrogen in their systems, which, in the low-pressure environment of their spacesuits, could generate bubbles in their blood streams, causing the bends. Before sending the first American woman astronaut out on an EVA, NASA had thought it wise to run a test to see if women were more susceptible to the bends than men; acccording to navy diving data and air force altitude chamber test data, there was a difference, and the percentage of greater suscep- tibility on the part of women was about the same as their greater per- centage of body fat, which might retain more nitrogen. The NASA test results bore out this theory, suggesting that Kathy would require a longer prebreathing period than Dave. A number of people and some women's groups immediately declared that the test and the decision were discriminatory. They argued that the women who took the test were all nurses, who are notorious for scrupulously reporting any physical symptom, whereas the men were all pilots, who are even more notorious for not reporting any medical problem that might in- terfere with their flying. NASA, whose last object in sending a woman on an EVA was to cause her injury, but who at the same time did not want to get into trouble with the women's movement, was in the sort of dilemma Sim Sup likes to think up. Dave saved the day by gallantly agreeing to start his own prebreathing early, at the same time as Kathy did.

After they had suited up and were lowered into the pool, I watched them on TV. Possibly because of their effervescent spirits, Dave had a greater tendency than usual to float to the surface and had asked for extra weights to be put in the pouches at his feet. (A little pre- breathing, I thought, would have taken care of those bubbles.) Kathy offered to help a little with what she called "down elevator"—hauling on his feet. Jon McBride, who would be monitoring the EVA from in- side the spacecraft, and whose own spirits were pretty high, said, "While you guys are out there, there's a big chunk of yellow ice on the port side." The flight controllers, the crew systems people, and the

EVA technicians all wanted to see if a crew member on EVA could easily hang over the port side of the spacecraft and kick away an icicle such as had appeared on 41-D. Dave could. Dave then began finding icicles all over the ship, including one on the orbital refueling system. "Kick it away," Jon told him. "Put it in your pocket."

They did an entire run-through of the EVA, which went flawlessly. Toward the end, a diver, Kitty Havens, who had been observing Kathy and Dave, came out of the water, got a towel, and sat next to Jim McBride, the instructor, who was watching the proceedings on television—there were four screens and two cameras. She had an earache and was clearly in a bad mood. "What I hate down there is turning around and finding a camera pointed at your hip or into your ear," she groused to the instructor as she toweled herself off. She watched Kathy Sullivan on television as she looked over Dave's shoulder. "She drops weights!" she said of Kathy Sullivan. "You put them on her, and five minutes later they're gone. I don't know how she does it!" She fingered a bruise on her arm. "You really have to stay out of the astronauts' way," she went on. "If they hit you, they have more mass than you, and they're covered with hardware; they can really hurt you."

At the end of the EVA, as Dave helped Kathy put the tools away, a diver came up behind him and floated a rubber alligator in front of his helmet. Dave let it bob there for a while, then he grabbed it and pushed it into the tool box. "I think it goes in lengthwise," he said, giving it a push. Clearly if they weren't launched soon, the earth's gravity would not be sufficient to keep them down on this planet, weights or no weights.

The Long Sim

*T*he three-day sim came ten days later. Ted Browder, who aimed all his training at the integrated sims, once told me that he directed it in particular at the three-day sim, which he regarded as the final exam. This was not because it was the hardest (indeed, for the astronauts, stand-alone training is generally tougher than integrated training), but because the long sim demonstrated more things in greater depth: because of its duration, more problems could build up and come to fruition. It was the only sim that let the astronauts experience the changing of shifts in the Control Room and the carrying over of problems from shift to shift, even from day to day. "In the other sims," Browder told me, "you go home at the end of it, and that's it; the problems go away. But in the three-day sim, you come back the next morning, and the problem is still there—maybe it's bigger and gives rise to some other problems." Because of their duration and the greater expense of running them, NASA managers trying to save money were forever trying to do away with them, but anyone who had ever been through one—astronauts, instructors, and flight controllers—was determined to keep them.

When I arrived, on the morning of the second day of the sim, the radiation budget satellite had been released just the day before (a day late, on the morning of Mission Day 2, the backup time) after a variety of ordeals, some of which had been planned at the script session I had attended and some of which had not. The umbilical hadn't released. While the crew had considered doing an EVA to fix it, its hydrazine began to leak. "No way I'm going to send my guys down in the bay," Crippen had said, as the instructors had known he would; they hadn't wanted to spend the time on the EVA. Instead, the ground had told the crew to use the arm to unplug the umbilical in the pull-the-entire-lamp technique. Then, when the radiation budget satellite was finally poised overhead on the arm, where the ground was supposed to check it out using its own communications system via the relay satellite, the relay satellite had mysteriously stopped transmitting—not because of a storm at White Sands, but because of a temporary loss of power there. The Remote Payload Operations Control Center at Goddard had had to figure out how to get their data by way of the NASA ground stations all over the world, which provided only intermittent communications. "Rainey and his men are very clever," Cox said with just a trace of bitterness, when he related all these troubles to me.

Shortly after the satellite was released, the spacecraft dropped down to its lower orbit for the shuttle-imaging radar, and the ground asked Jon McBride to realign the inertial-measurement units. Because of a combination of errors, the realignment was way off: the ground had inserted some new navigational coordinates into the computer but had neglected to inform Jon, and Jon had punched in some old numbers despite a discrepancy that should have told him matters were not right. As mistakes go, it was a minor one, easily corrected. There have been major ones in the course of integrated training for other crews. During an ascent sim, when one pilot reached over to turn off a main engine, he turned off all three by mistake. On another sim, a commander turned off all three inertial measurement units, so that the spacecraft couldn't sense motion; consequently the thrusters kept firing and firing, so that the spacecraft spun faster and faster. The shuttle, according to many people familiar with it, is not user-friendly.

This time, Crippen saved the situation by tapping the right information, consistent with the new coordinates, into the computer. The integrated instructors had as much admiration for Crippen's quickness and skill as the stand-alone team had. As was the case in the stand-alones, he was quicker than most of the flight controllers in the integrateds at scanning the telemetry data and spotting a problem in a column of numbers—say, a rise in pressure in an oxygen tank which might cause a rupture. He often solved such problems before the flight controllers were aware of them. He is, of course, particularly adept with computer problems. Cox told me, "He drives the data-processing systems people nuts. If there's a computer problem, Crip will come up with the answer while our computer guys are still wading through the procedures. They'll agree he's right, but they won't know how he got there. It'll take them two shifts to find out. With most crews, they'll see an alarm, and at the same time we'll see it, too. The crew will ask us to evaluate it. It'll take us three or four minutes to answer. Typically with a Crippen crew, we'll see a malf, we'll start figuring it out, and he'll have the answer before we do." And Susan Creasey, who is a Sim Sup on a different shift from Rainey's, said, "Often, when we do these integrated sims, Crippen will solve all the problems, and that's bad for flight controllers. A Crippen crew can spoil those guys—make them think sitting at a console is a breeze—and then they'll get a green crew and they won't know what to expect."

Crippen was learning to restrain himself in the integrated sims, just as he had done in the stand-alones. Now, though, he had to restrain his crew as well, something that was no easier for Crippen than restraining his team was for Browder. Early in the second day of the long sim, just before I got there, McBride noticed a very slow leak building up in a helium tank for pressurizing the thruster system. He was clearly champing at the bit to report his discovery to the ground, but Crippen stopped him—"Let the ground find it for themselves," he said. "That way, they'll get something out of it, too." Jon was a little sore; the situation was more than a rookie pilot who had recently become proficient should be asked to bear. Later, Crippen admitted to me that he and his crew did indeed play 'possum with a problem, lying low to let the ground find it, though he added modestly that no one should assume that *all* the problems that slipped by him were purposely allowed to do so. Of course, in a real mission, neither Crippen nor McBride would ever hold back essential information.

In the lower orbit, the imaging radar antenna with its two side panels had been opened, but there was something wrong with the ground's ability to control the angle at which it was tilted, so that now it could not be aimed at its targets to one side or the other of the ground path. The astronauts had to do this.

"Where is the radar antenna?" the capsule communicator—an astronaut named David Hilmers—asked.

"The antenna is in the payload bay," David Leestma said.

"Thank you very much, Dave," the capsule communicator said, with a touch of sarcasm.

"It is at forty degrees," Kathy Sullivan said, telling Hilmers what he wanted to know.

Rainey broke in on the loop to Cox to say that the ground should avoid imprecise questions, such as had elicited the imprecise answer from Leestma—though I found out later that Leestma knew perfectly well what Hilmers wanted, and was helping instruct the ground in his own way.

The astronauts were in fact unable to open the antenna all the way (to sixty-one degrees) and lock it in place, because of the failure of a microswitch. The ground began radioing up small attitude changes for the orbiter itself, tilting the entire craft from side to side of the ground track, so that it could help the antenna reach the higher angles. But this solution, once it proved effective, was not allowed to work for long. A computer that operated the Ku-band radio antenna (the black dish at the rear of the cargo bay through which data was transmitted to the relay satellite and thence to the ground at White Sands) suddenly broke, so that the dish couldn't be aimed at the satellite automatically; the aiming had to be done by orienting the entire spacecraft. Because it was now impossible for the spacecraft to aim the imaging radar antenna at the ground targets and the Ku-band radio antenna at the relay satellite at the same time, it meant that the radar scientists had to decide from which targets they were going to take data (and store it on tape) and which targets they were going to pass up, so they could dump the tapes to the relay satellite or to one or another of the ground stations. Jim Freehling, the communications instructor, a tall, blond young man with a crew cut, was pleased that he had got the problem in; Rainey tended to veto his more dramatic efforts because communications problems were likely to cut off the data the flight controllers needed to solve all the other problems—and a sim with no data was no sim at all. "It's sort of Catch-22," Freehling said. "If you want to train people in major comm breakdowns, they don't get any training." He consoled himself with the fact that communications were so reliable that major comm problems in orbit are very rare.

Keith Todd told Rainey he was ready for what he called his "massive malf," by which he meant failing the single-joint mode of the arm. This would have to be done joint by joint in the simulator. "It'll wear out the guy with the light pen," Todd said, meaning Basham.

Rainey said, "Yes, and if it brings the computers down, I'll be mad." There had been several computer crashes the day before, and no one wanted any more.

A pump that circulated freon to cool the instruments in the payload bay was acting up. John Cox suggested to the controller in charge of the electrical and environmental systems that the astronauts recycle the pump's circuit breaker (that is, turn its switch on and off) four or five times in quick succession and then hold it in, in the "on" position. The electrical and environmental controller said he doubted if it would work. Cox said it had worked once before, and sent word to McBride in the spacecraft to try it. Jon gave it a try, but got no joy. Cox sent word to him to try again. After several tries, it worked. Cox looked pleased with himself. I found myself wondering how many of the problems I was watching would occur on the mission itself, either in their present or in some other form—or were the sims simply exercises with no real bearing on the mission?

I joined Cox at his console in the Control Room. He was talking to the flight controller in charge of the arm, who wasn't sure the single-joint mode—which Todd had managed to break without crashing the computer—could be fixed, and Cox was telling him severely that if it wasn't fixed in time to berth the arm before the spacecraft returned to earth, it would have to be blown off the vehicle with explosives. The controller looked unhappy. After he had gone, Cox told me he suspected that the arm people would find a solution. "If something breaks on board, they're so damn smart down there that they can figure anything out, given enough time—and that's where we earn our bread," he said. (The night shift of flight controllers, on duty from midnight to 8 a.m., normally worked on problems that came up during the day, and hence it was known as the planning shift; they would work on the arm.) Another flight controller came up and confessed to Cox that he had screwed up some problem, and was sorry. Cox told him to forget it; screwing up was what sims were for.

At about three-thirty in the afternoon—half an hour before the change of shift at four—the new team of flight controllers joined the old team at their consoles in the control room, and the new team of integrated instructors joined the old team at their consoles in the Simulation Control Area, so that the outgoing people could brief the incoming ones about what was going on. Jim Freehling was filling in his replacement about the broken Ku-band radio antenna, and how the radar scientists had to keep rewriting the timeline for their data takes and transmissions. Jack James was telling his replacement about the slow leak in the helium that pressurized the hydrazine for some thrusters—the leak that Crippen and McBride were keeping quiet about. It had been put in ten hours ago—two hours before James himself had come in that morning. The leak was being quietly handed from shift to shift, like a secret message around a dinner table.

*T*he next morning Rainey's team was beginning to think about bringing the astronauts home just before the end of the sim. They were going to do this by causing a leak in the cabin pressure—a slow, inexorable drop in pressure, as I had seen once during the stand-alones. At the script-writing session, Jack James had told me, "That's about the only thing that will cause an emergency de-orbit these days. A fuel leak would either be too slow or be catastrophic; either way, it wouldn't do the trick. But a cabin leak has a predictable curve the flight controllers can plot, and estimate how much time they have before the spacecraft has to be on the ground. That way, they can pick the best emergency landing site." Cabin leaks are a favorite way of ending sims. James, who always wanted to be a step ahead of the flight controllers, wanted to know what landing site they were likely to pick, even before they knew there was a leak. There was a table of the different emergency landing opportunities available on each orbit, but James had left his copy of it back on his desk in Building 4. Accordingly, he called the flight controller in charge of trajectories, the flight dynamics officer, and asked him what the landing opportunities were around noon. Rainey, who was listening in on the same loop, shook his head. "I think you tipped our hand to them," he said. "There's a lot of chatter about the question in the back room. I suspect they're on to us. But it's near the end of the sim, so they ought to suspect *something*. For them, it's in the rumor stage." Crippen had once told me that both astronauts and flight controllers liked to try to psych out their instructors, and that this was usually a good thing because it was an exercise in foreseeing what might happen. On a real mission, though, I wondered how often Fate would insert a major malf, with the intent of bringing the astronauts home in time for lunch or dinner.

All the instructors were now turning their attention to making the emergency de-orbit, and the return to earth, as rough as possible. Carl Keppler, the silver-haired electrical and environmental systems instructor—who would have the job of causing the cabin leak—was listening intently on one of the loops. "I can listen to them, and I can see all the things they're thinking about, and then I can zap them with some other things they haven't thought about," he said. Clearly, it gave him the same godlike feeling it gave Rainey. He wanted to get the most mileage out of the cabin leak, because all by itself it would be uninteresting. At the moment he was preparing an embellishment, a failure in a pump in one of the auxiliary power units that control the flaps for flying in the atmosphere and for lowering the wheels. There was some honor, though, even among mind-reading instructors: because there were two redundant auxiliary power systems, there was only a fifty-fifty chance that the astronauts would turn on the one he had booby-trapped. "I could fail the other system, but that really would be forcing them in an unfair way," he said. Reality was random; either they would pick his power unit, or they wouldn't. James, I had noticed, was not so squeamish about shoving people into traps when they didn't fall in—nor, for that matter, it turned out, was Keppler, in the final analysis. He had no compunctions about doing something else to them if they picked the wrong power unit. "I may drop their cabin temperature on the way down, or I may do something to the astronauts' oxygen helmets," Keppler said. "It's a long, dead time down to the ground and we have to keep giving them something. We have all this sim equipment—it's a waste of time and money not to use it. I have to keep listening to the back room to see what they're thinking about and to keep getting ideas for how to play it." He arranged with Randy Barckholtz in the mission simulator that if the astronauts picked the right power unit, the pump would explode with a really satisfying thump.

Because it was not always certain which of two or more alternative pathways astronauts or flight controllers would pick in solving a problem (this was one of the factors that made sims unpredictable and realistic), it was quite permissible for the instructors not just to booby-trap a single sinuous trail, but to mine all the trails the astronauts or flight controllers might conceivably take—though not, of course, with the same bombs. Steve Hamm, the computer instructor, was setting things up so that whichever of two master units in the entry computer the astronauts selected, they would be sorry, though for different reasons: if they chose Master Unit 1, they would have no communications with the ground; if they chose Master Unit 2, they would not have any information about the environmental system in the cabin—including information about Keppler's temperature problem, which would be minor, but, more importantly, about his cabin leak, which would be major enough to bring them home. Jim Freehling, who was helping Hamm and who told me about the problem, said, "So it's like a chess game, when your knight is forking two of your opponent's castles. It's got all the complex rules of chess, as well as the unknown factor of what the personalities will do." No one had much time to savor the niceties of chess moves, for just then Crippen reported that the freon system for the payloads in the bay had given up: the pump had finally stopped, despite (or because of) all Cox's recycling of its switch yesterday.

Keith Todd, at Rainey's urging and against his own better judgment, had restored to the astronauts the full use of the arm's single-joint operation, so that they would be able to berth it when the time came to bring it home. The flight controllers Cox had exhorted the day before had not come up with a fix, and Rainey did not want to penalize them—Rainey, I realized, sometimes had to exert the same moderating influence on his team that Browder did, and his team members were sometimes just as ingenious at circumventing him. Todd, who did not like to see his massive malf simply evaporate, told me he was going to put a sensor failure in the berthing mechanism for the arm, so that when the astronauts stowed it, neither they nor the ground would get confirmation that it was secure. According to a mission rule, it had to be latched in place before it could be brought home, and so the ground would have to consider once again whether to jettison it—they would, in fact, be more or less where they would have been had Rainey not asked Todd to restore the arm's single-joint operation, and that was right where Todd wanted them.

Keppler was listening with a broad grin on his face to the electrical and environmental controller's back room loop, and also to the air-to-ground. Clearly, many people were falling into a trap. The astronauts were hastily powering down equipment, and the electrical and environmental controller was telling them what to shut off next. "I've busted one of the two main freon loops for cooling the entire spacecraft," Keppler explained. "A meteor has hit the plumbing for one of the two big radiators inside the cargo-bay doors, and they've lost half their cooling for all the equipment in the spacecraft." He wasn't too sure about the meteor—the real cause, he said, was simply that he had wanted to do it. Ed Rainey told him that he hoped he had figured things so that the astronauts would make it to the ground before the fuel cells that generated electricity—highly sensitive to temperature—overheated and failed. Keppler said of course he had—though I wasn't sure how confident he was, for a moment later he said, "Fooling around with the cooling loops can be a tricky business." There were two additional systems that supplemented the radiators for dispelling heat (the flash evaporators and the ammonia boiler), but they were not very efficient; and even though equipment could be powered down, to help with the cooling, still a lot of it would be needed for entry.

Keppler's big moment was now at hand. He told Randy Barckholtz over in the simulator that it was time to insert the cabin pressure leak, which was theoretically due to a faulty seal around some windows, or maybe it was due to a small meteor; meteors were evidently something of a specialty with Keppler. He was putting the leak in ten minutes earlier than originally planned because the spacecraft would be out of contact with the relay satellite at the originally scheduled time, and by the time the spacecraft was in contact again and the flight controllers were aware of the problem, they would have too little time to work on it. Because the leak was going in earlier, Keppler had had to reduce its size, and in doing that he had made it too small: instead of requiring a de-orbit burn in two hours' time, he had reduced the leak so much that it wouldn't require one for five hours. The technicians in the trajectory back room had already figured the time and the emergency landing site: it would be at the Northrup strip in New Mexico, at about 5 p.m. With the debriefing, nobody would get out until 9 p.m., and the post-sim beer bust, a tradition after all long sims, was scheduled for 5 p.m.

"The leak's too slow!" Rainey said to Keppler. "Can't we speed it up?"

Keppler made a quick phone call to Randy Barckholtz. Moments later, the electrical and environmental controller, who was in charge of the cabin environment, reported an increased drop in cabin pressure. It was automatically met by a sudden surge of air into the cabin from the oxygen and nitrogen tanks. If the pressure in the cabin dropped below eight pounds per square inch, the astronauts would don helmets and breathe oxygen directly; if they landed before it dropped below three pounds per square inch, they probably would not get the bends. The electrical and environmental back room recomputed the length of time allowable before the de-orbit burn—it had to be done within two hours. The trajectory back room announced that the spacecraft would land in Honolulu at 2:30 p.m., just as they had tried to do after the cabin leak in the stand-alones, when they had ditched near the surfers. All the instructors cheered because it was clear the sim and the subsequent debriefing would be over before the beer party, and there was renewed activity on the loops.

The astronauts set about their de-orbit preparations. Sally stowed the arm but could get no confirmation that it was locked in place, because of Keith Todd's new nit with the berthing sensor. He listened on the loop, a smug look gradually spreading across his face. "They're trying to decide whether or not to jettison," he said. "Myself, I wouldn't jettison."

"But you're the one who put the problem in," Rainey objected, "and you *know* the arm is O.K." He did not like this nit, which he had not approved in the first place.

The smug look vanished. Todd protested that he wouldn't jettison in any case.

As the astronauts turned on more and more equipment in preparation for the de-orbit burn, the cabin—cooled now by only one freon loop—began to heat up, and Crippen and McBride, on instructions from the ground, had to turn off other equipment that was not necessary for entry. Still the cabin did not cool down enough. Keppler looked worried; Rainey would not appreciate it if he brought down the sim. Before a normal entry, when there is plenty of time before the de-orbit burn, the payload bay is turned away from the sun to cool (or, as space engineers call it, to "cold-soak") the bay and the silver radiator panels. That way, even when the payload-bay doors are shut for entry, the radiators can still get rid of enough heat so that the spacecraft can get to the ground safely. Now there was no time for the cold soak. It was not clear whether Keppler had taken that into account when he had hurled his meteor.

Steve Hamm was talking animatedly over his headset with Shannon O'Roark in the mission simulator. "Crippen did *that?*" he said with a pleased smile on his face. I asked him what it was that Crippen had done. He had brought on line several of the general purpose computers to perform the program known as Entry Operations—including one computer, number 5, which very early in the sim had been retired because it had lost its ability to control strings of equipment and therefore should not have been brought back on line at all. Had Crippen, of all people, made a horrendous computer error that would rank in sim history along with that of the astronaut who had turned off all three main engines during ascent? Hamm wondered this and called Shannon, who of course was in close touch with the astronauts and who told him that Crippen was well aware of the situation—he had simply done what he had been told to do by the capsule communicator, who was passing on instructions from flight controllers. When she had asked him about it, Crippen had said, "I turned it on because the ground didn't tell me not to." Hamm said to me, "It's his way of helping to train the flight controllers. And it's a good point. There are only five astronauts in the crew, and they shouldn't have to think of everything." Crippen, I knew, often played a passive, 'possumlike role in training by not solving problems so that the flight controllers could get practice; this was the first time I had heard of his taking the active role of agent provocateur.

The spacecraft was out of contact with the relay satellite now; the burn would occur when they were in contact again. While they were out of contact, the astronauts had shut the payload-bay doors—Sally Ride had visually inspected the arm and had decided that it was locked in place. Although the astronauts did not normally make decisions while they were out of contact, in preparing for emergency de-orbits they had to.

Jack James, the propulsion man, was emplacing a little squib to go off at the start of the burn, which was now about half an hour away. One difference between the stand-alones and the integrateds was that the greater number of instructors resulted in a greater mass of ideas, many more of which could be done because of the great number of flight controllers to solve them; there was little worry about overtaxing the astronauts. If the stand-alones were aesthetically purer and the relationships simpler and more direct, then the integrateds were richer—and as the end approached, it was like watching the finale of, say, *The Mikado,* with a great variety of subplots coming to their conclusion simultaneously in a happy crescendo of malfs, glitches, nits, anomalies, and bites. "I'm going to kill the right orbital maneuvering system engine when the burn begins," James said. "The mission rules call for shutting down and staying in orbit, to see if the trouble can be fixed. This time, they can't remain in orbit because of the cabin leak, so they will have to do the de-orbit with a one-engine burn. I want to see if they follow the knee-jerk rules and shut down the burn. I'm sure Crippen wouldn't do such a thing, but maybe some yo-yo on the ground will try to tell him to."

Indeed, a lot of things were coming together at once, sometimes in unanticipated ways. Crippen now gave McBride the go-ahead to reveal the helium leak in the thruster system, which he had found the day before: if the flight controllers hadn't found it yet, they never would. Jon, unable to contain himself, burst out, "Can anyone down there tell me anything about this leak that I've been tracking for the last thirty hours?" McBride himself, of course, was becoming wonderfully skillful, and it was hard for him now to see how a back room full of technicians looking for such things on their computer screens could have missed it all that time. Jack James, wreathed in smiles, tuned in on the propulsion controllers' back room. "They're jumping all over it," he reported. Then he laughed. "Cox is worried that the leak is going to build up and cause an alarm during entry. That's a possibility I hadn't thought of."

Crippen was now reporting the failure of computer number 5 with all the innocence of a rookie, and the ground belatedly told him to switch it off and turn on another.

The flight controllers, who were not aware that the astronauts had shut the doors when they were out of radio contact, were still arguing about whether or not to jettison the arm. Keith Todd was looking incredulous. "The flight director doesn't know the doors are closed!" he said. "The arm people should have told him—it's their system." The spacecraft had been back in contact for some time, and all the flight controllers had to do was look at a code number on a telemetry chart, Todd said. "They've decided against jettisoning; they say the arm looks straight, and they're go for door closure. At least they came to the right decision. Better late than never." He listened some more and reported that the ground now had found out that the doors were closed. They had apparently been prevented from knowing earlier because the defective computer had not allowed them to call up the telemetry about the doors; now that computer number 5 had been replaced, they instantly saw the true situation. "So my problem played into that problem—I hadn't known," Todd said. He thought less harshly of the flight controllers.

The moment for the burn arrived: the only way the instructors or flight controllers could tell was, first, by watching the numbers on a countdown clock reach zero and then begin going the other way, and then by watching other sets of figures for the fuel pressure drop. When the right maneuvering system engine failed, no flight controller was dumb enough to suggest turning off the burn. "They didn't take the bait!" Rainey said to James. McBride remembered to cross-feed at the right moment, and afterward when he dumped the remaining maneuvering system fuel to shift the craft's center of gravity for entry, he did not dump the thruster fuel as well.

Everyone was in high spirits. On the way down, there was a burst of minor nits, easily solved. Keppler caused an alarm by failing a smoke detector, and James put a leak in part of the thruster system, so that hydrazine dribbled into the cargo bay. As the craft hit the upper atmosphere, James said joyously if inaccurately, "They're still trying to figure out if the doors are closed," and Todd echoed, "They're still wondering whether to jettison the arm!" If either of those slanderous statements had been true, the spacecraft would have burned up.

Rainey discovered that a switch controlling the orbiter's ultra-high-frequency radio system, which would allow the spacecraft to talk with the tower at Honolulu, had inadvertently been turned off, so that in the last minutes of the flight the astronauts would have no communication with the ground, either at the airport or at Houston. Rainey passed the word to Browder to tell Crippen to turn it on; it was not part of anyone's script to have no communications at this point.

Keppler reported uneasily that the temperature of the radiators, now closed inside the cargo-bay doors and not cool enough to begin with, was 125 degrees Fahrenheit—much too high. He was afraid the craft would lose all three of its fuel cells, and therefore its entire capacity to generate electricity, before the landing. "They're working fine now, but when they go, they go fast," he said unhappily.

"If they fail, you can still show up at the beer bust—for your lynching," Rainey told him. A spacecraft had never been lost in a three-day sim before.

"Three minutes to touchdown," Keppler said a little later. "We'll make it."

They did. Instantly the crew had to do an emergency power-down. This was followed by an emergency egress, followed by lunch, everyone staying at their consoles or crew stations. "If the orbiter worked as badly in space as it does in sims," Hamm said, biting into a thick sandwich, "we wouldn't use it." He added, "But if everyone learns on sims, the flight will be a big yawn. And if the flight is boring, we know we've done our jobs."

The debriefing began right away at a little after two, and over the next three hours, under the chairmanship of John Cox, they analyzed every problem that had come up during the long sim and discussed how it might or might not have been solved better. The instructors laughed when the flight controller for the radiation budget satellite discussed all the false trails she and the remote control center at Goddard had followed when they had been unable to detach the umbilical. Cox chided the trajectory people for taking so long, after the satellite was deployed, to get down to the lower orbit. Cox took the astronauts to task for having certain switches in the wrong position. Crippen suggested that, after one of the computer crashes on the first day, perhaps a technician resetting the simulator had caused the error. Cox said that that had not been the case. "O.K., so we goofed up," Crippen admitted grudgingly. "But if we ever leave the switches that way, I'd appreciate hearing about it from the ground."

When each flight controller had spoken and the debriefing was almost over, Cox turned the session over to Rainey with the words, "O.K., your nickel, Sim Sup." This is the traditional ending of most integrated sims.

Rainey, who hadn't spoken much yet, was expected to discuss any problem that might not have been touched on or to correct any misunderstandings any of the flight controllers might have had about any of the problems. The only loose end he could think of was the radio switch that had been off. Crippen said he didn't know how it had happened.

"It was probably done by a payload specialist," Steve Hamm said to me. "Stuff him in the airlock!" The integrated instructors were as chauvinistic as the stand-alone instructors. Payload specialists were generally regarded as such ineffable rookies that I wondered if they would ever gain veteran status, even after they had flown in space.

"Thanks for simming with us," Cox said to the crew, by way of ending.

Signs had been posted all over Mission Control inviting everyone who had helped with the long sim to the beer bust—the astronauts' party to thank the flight controllers and everyone else who had helped with the mission. The party was held under a pavilion in a field at the back side of the Johnson Space Center; under the pavilion, which was open on all four sides, were a half-dozen big, bright red kegs of tap beer. It flowed from hoses into cups, and flight controllers, instructors, and astronauts took turns pumping the handles. A couple of hundred people were there, and under the roof the noise was deafening. I bumped into Ted Browder, who was wearing his usual light blue T-shirt and who said he was very happy with the way the long sim had gone. The astronauts had passed their final exam with flying colors, Ted said. "The way you can gauge a crew's performance is by how well it does in relation to the ground. Some crews get the answer to a problem before the ground. Some crews are just equal to the ground. And a few crews are behind the ground. This crew was almost always ahead of the ground, and so I'm sure you can see why I'm ecstatically pleased by their performance. I think they're ready right now. I'd have no qualms about putting them in the orbiter today. From now on, their training is just a matter of fine tuning." There were only about two weeks until launch.

Kathy Sullivan was autographing a shuttle-imaging radar team T-shirt for one of the radar scientists, using a red magic marker. Sally Ride was autographing a mission document for a flight controller. Jon McBride was regaling a small group about the helium leak when a friend came up and handed him an audiocassette of a country-Western song the friend had composed and sung, entitled "The Song of Jon McBride." Jon promised to take it to orbit and play it there.

On Friday, September 21, a couple of days after the end of the long sim, Kathy and Dave were in the WET-F, practicing a complete run-through of their EVA, this time with the flight controllers' loop plugged into their own; the flight controllers and the astronauts were simming the entire fifth day of the mission, and the EVA was part of it. So that Dave and Kathy could help simulate malfunctions that might come up on a real EVA, divers held up signs in front of their helmets that bore messages such as "YOUR SUIT IS OVERHEATING" or "YOUR TOOL HOLDER IS JAMMED"; then Dave or Kathy would put on a little act—complaining about the stuffiness or saying "Rats!" a few times—for the benefit of the flight controllers and the spacesuit technicians in their back room at Mission Control. The low point of the sim came when Dave's flashlight actually broke in half: he had one piece in each hand. "I can still use it," he said. "I'll use the mirror and bounce my helmet light off of it." Kathy volunteered to let him use *her* as a flashlight—presumably he would aim her head first at the orbital refueling system, with the lights on either side of her helmet turned on. The high point of the EVA was a simulated telephone call from the president, with Jack James taking the part of Ronald Reagan, who would be on a campaign trip during the mission and hence was calling from an American Legion hall in Des Moines. James not only had a good dramatic sense, but he was also a good mimic. In the president's voice, he asked Dave and Kathy if they were refueling a real satellite; called Kathy "the first free woman in space"; and invited them all for dinner at the White House when they came home. "Does that mean we all go to Jack's house when the sim is over?" Kathy asked, stepping for a moment out of her role. There was general laughter on all the loops, and nothing further was heard from James. I hadn't quite realized before what a closely knit group all these people are who prepare missions—the seventy-odd astronauts and one hundred sixty or so instructors with offices in Building 4, the few hundred flight controllers across the duck pond, and the engineers who are most closely associated with them add up to well under a thousand people, the size of a small college; and as in any such small institution where people work hard and perhaps play hard together, there is a tight sense of community.

Countdown

The last two weeks went by very fast. The long sim was over on Wednesday, September 19; the launch was sixteen days later, on Friday, October 5. If NASA was deliberately trying to keep the astronauts occupied so that they wouldn't have time to worry about what might go wrong, it couldn't have planned the last few weeks better. Sally Ride told me later, "In the simulators, they gave us malf after malf after malf in a way that is completely unrealistic—coming out of those sessions, you feel completely beat down." Dave Leestma said that toward the end he was tired. One member of the crew talked of feeling "numb." I never found any evidence, though, that NASA was deliberately trying to condition the astronauts for a dangerous assignment by working them into a state of numbness ahead of time. Given the universal tendency for schedules to slip and things to slide to the last possible minute, the final weeks before any mission, like the preparations for any expedition anywhere, are apt to be hectic. And Crippen and the crew did not strike me as being in need of anesthesia—unless it was to make them less excited. They wanted to go into space more than anything else in the world. To the extent that they might have had some last-minute butterflies, the best antidote at the time of 41-G was simply living and working at the Space Center, for everyone there—instructors, flight controllers, astronauts, and engineers—while being fully aware of the dangers, was very upbeat about space flight. In addition, the sims had given them the added confidence that they would be able to handle almost any problem that came up.

On Thursday, September 20, the day after the long sim, Crippen had a photography briefing in the first part of the morning, and Kathy Sullivan did a final check of the spacesuit she would be wearing on her EVA. From ten to noon, all the astronauts had a session with scientists who told them what to look for on the earth. On Thursday afternoon,

Jon McBride flew ascents in a simulator, and Dave Leestma studied potential malfunctions that might crop up with the orbital refueling system.

On Friday, September 21, all the astronauts participated in the integrated sim of Day 5 of the mission—the day of the EVA.

On Saturday, September 22, Crippen and McBride flew in the Gulfstream to Edwards Air Force Base in California and practiced night landings.

Sunday, September 23, was a day off—the crew's first in several weeks. It was also their last day at their own homes. On Sunday night, the entire crew moved into a trailer at the back of the Space Center, where they would start a quarantine period during which they would see face-to-face only a very few people on essential business; the purpose of the quarantine was not just to shield them from catching colds, but to give them a little rest.

On Monday, September 24, they did integrated sims of the ascent in the morning, and in the afternoon they did an integrated sim of Mission Day 6, with particular emphasis on gathering data with the instruments in the payload bay.

On Tuesday, September 25, they spent the day in their office, catching up on paperwork.

On Wednesday, September 27, they had their preflight physical examinations, including a dental check. (No one wanted a toothache in space; the Skylab astronauts, who were up for as much as three months, knew how to extract teeth and had the equipment to do it. They could also do emergency appendectomies. Crip, Jon, Sally, Dave, and Kathy did have these skills.) On Thursday afternoon, they spent some time with a public affairs officer to discuss the television they would be transmitting from space; then they had a final review of the flight data file—the crew activity plan and all the checklists and procedures, over ninety pounds of them, that would be aboard the *Challenger*.

On Friday, September 28, they spent most of the day in integrated sims of preparations for de-orbit—closing the payload-bay doors, firing the maneuvering system engines. After the sim was over, the crew met for an hour with John Cox and the flight directors of the two other shifts of flight controllers for 41-G to discuss final plans for the mission.

The weekend of Saturday, September 29, and Sunday, September 30, at Crippen's insistence, was entirely free.

On Monday, October 1, the astronauts spent the morning doing integrated sims of entry—their last practice of the return to earth. (The new computer loads for the simulator continued to arrive late, so that Crippen almost didn't get to try out the latest updates for the return to earth.) That afternoon, Kathy and Dave had their last EVA session in the WET-F. That night, the crew invited Browder and the team over to their quarantine trailer for snacks and beer.

On Tuesday, October 2, in the morning, the astronauts did their last simulation of all—integrated sims of the ascent. Training traditionally ends with ascents. There were four runs in all. The first was a return-to-launch-site abort. The second run was supposed to be an abort to orbit with one engine stuck in the bucket. They in fact couldn't make it to orbit, so the abort quickly had to be turned into an RTLS. "We weren't sure they could make it back all right, but they did," Browder told me. "Even this late in the training, we are still learning about their, and the *Challenger*'s, capabilities." The last two runs were nominal—something almost unheard of in a sim. "We don't want to end on a sour note," Browder said. "We want them to get a look at what things might really be like on Friday morning, in the real world." It was also a way of wishing the crew luck.

The team and the crew said goodbye to each other in the simulators. The team would be driving to Florida to see the crew fly off to space; they all told each other to drive carefully.

Tuesday afternoon, the crew flew to the Cape in T-38's. They held a short press conference at the airstrip after their arrival and then went to the crew quarters, an apartment in a big building adjacent to the Kennedy Space Center's headquarters. They were, of course, still in quarantine. They went to bed early, at between 6 and 7 p.m. For the last week or so, they had been backing up their bedtime and getting up earlier, because they would have to get up on Friday morning at 2:30 a.m. to be ready for the launch at 7:03. They would continue to get up at 2:30 a.m. throughout the mission.

On the morning of Wednesday, October 3, the astronauts were awakened between 3 and 5 a.m. Crippen woke the earliest because he was flying approaches and landings at the Kennedy airstrip in the Gulfstream. The rest of the astronauts did fit-checks of their equipment.

For lunch on Wednesday, they had a cookout at the beach with their families and some friends. Browder and the team were invited, but because they were late in arriving from Houston, they decided not to barge in on the astronauts and their families. After an afternoon on

Sally, Crip, Kathy, Dave, and Jon depart for the Kennedy Space Center a few days before launch.

The landing strip at the Kennedy Space Center, as it might be seen by an orbiter approaching from space. The Vehicle Assembly Building is at the top left of center.

the beach, the astronauts went home to bed. First, though, they telephoned the team at the Wakula Hotel.

"Bring some cold beer to Kennedy for us when we land," Crippen told the team.

"But you keep going to Edwards," Browder said.

"You can bring the cold beer to Ellington," Crippen said. They would catch up with the beer there. Wherever they landed, the astronauts would fly home by way of Ellington.

The team promised they would do just that.

The astronauts were in bed by 7 p.m.

On the morning of Thursday, October 4, the astronauts were awakened again between 3 and 5:30 a.m. Jon McBride practiced approaches and landings in the Gulfstream. At 10:30 a.m., they were briefed on the status of the *Challenger* and its cargo and told of any changes that had been made since they were last at the Cape three weeks before, for the Terminal Countdown and Demonstration Test. They were also briefed on the next day's weather. The briefings were done by telephone conference; the closer to launch, the stricter the quarantine.

After lunch at about 11 a.m., they had a final review of the flight data file, they discussed the procedures for entering the *Challenger,* and Crippen and McBride both practiced landings in T-38's.

Dinner was at 3 p.m. and the astronauts were in bed at 5 p.m.

On Friday, October 5, the astronauts were awakened at 2:30 a.m. by a technician, Frank Newman, who often performs this service. "I woke them all up," he told me. "It made me feel good to make all that racket." A couple of the astronauts were already up—Sally, for one. They were all excited and keyed up, particularly Kathy.

At 3 a.m., they had breakfast, followed by a telephone conference with Gerald D. Griffin, director of the Johnson Space Center in Houston, who wished them luck. The two payload specialists were wired with biomedical gear.

At 4:10, they started for the pad in a brown minibus called the Astrovan, bought recently by NASA to accommodate the new, larger crews.

At 4:35, they arrived at the pad.

At 4:45, they began entering the orbiter.

At 7:03 a.m. (6:03 in Houston), the crew was launched. The team watched from the bleachers. "We felt about them as if they were family—and I think they felt that way about us," Browder told me when I saw him later that morning. "I care about what happens to them. I want them to do well. As I said earlier, when they fly, a piece of us will be flying with them. And I want *it* to do well!"

A Step Up

To the flight controllers in Houston, who took charge of the mission the moment the spacecraft cleared the launch tower, the flight might almost have been another simulation. The closed-in, windowless character of the Mission Operations Control Center, which, during simulations, had caused the participants to feel they were engaging in a real mission, now created the opposite illusion: that the mission was a sim. Most of the controllers were sitting at the same consoles, or in the same back rooms, as they had during the sims; there was almost nothing to tell them that the astronauts weren't still across the duck pond, in the mission simulators, instead of orbiting a couple of hundred miles overhead. As in the sims, there were the same long periods of silence, when the astronauts were out of touch with the ground. At the front of the Control Room was the familiar map of the earth with the same silhouette of the shuttle moving slowly along its orbit. To the flight controllers, the sims and the flight were almost a single, continuous, unified experience. "The space flight to me is just like a simulation," William Holmberg, the flight activity officer told me; he now sat in the Control Room a couple of consoles to the right of the flight director, where he would oversee the execution of the crew activity plan he had developed. "It's hard for me to believe there are people up there, flying around the earth while I'm sitting at my console—it's so like a sim!"

Since I had seen the launch, and therefore experienced firsthand the gigantic controlled explosion whose force had put the spacecraft into orbit around the earth, I had less difficulty believing that the little projected silhouette of the shuttle inching across the big map measured real motion. To someone who had observed the training, the most interesting question was to what extent the sims had been accurate in representing the mission, and how much the mission would be like the sims.

About the only ways the flight controllers could tell that the sim world and the real world had become one—that what appeared to be an imitation was in fact the genuine article—were that all the men were wearing jackets and ties, that the problems were not as frequent or as serious as usual, and that the curtains on the window to the simulation control area were now closed. Inside the simulation area, instead of Rainey and his team, whose roles in the mission were now over, were Ted Browder and some of the stand-alone instructors. (Another group of instructors had covered for them until they got back from the Cape.) Since Browder and his team knew the astronauts and their quirks better than anyone else, they had to be available to advise the flight director on any matter concerning the crew. "We know what the crew might or might not be able to do," Browder told me. "We can say whether or not the crew will understand some new procedure Mission Control wants them to do. Could Crip or Jon handle a certain type of abort? Could Sally and Dave handle an entry by themselves, if they had to? We're the resident authorities on the crew." They would also have to be on hand to conduct simulations in support of the mission, as had been done for previous missions. Consequently Browder and his team in effect became flight controllers, and the simulation area was their back room. "Our role in the mission isn't over until wheel-stop," Browder told me. "On the training charts, the last block is the flight. The flight is the ultimate sim."

The astronauts, though, were under no illusions that they were participating in another sim. Although they recognized that the flight was firmly rooted in the training, they could never confuse the two. To them, the relationship was not a smooth continuity, but rather a profound discontinuity. "The training certainly does lead to something, and is part of the experience—but then there's this big step," Sally said when I saw her afterward. And Dave used almost the same words: "The training is obviously building up to something, but then there's this neat thing at the end—a step up." While to the flight controllers and instructors the flight and the training is almost a single, unbroken, unified experience, to the crew the relationship is more like that of a jewel to its setting.

The astronauts entered the spacecraft—each one in turn helped through the hatch by a technician—with those who had the farthest to go clambering in first: Crippen and McBride, in the forward cockpit seats, led the way, followed by Sally and Kathy, who would be in the rear cockpit, in seats that would be folded up and put away in orbit;

then the two payload specialists, who would be on the far side of the middeck in folding seats, and finally Dave Leestma, whose folding seat was nearest the hatch, where all the others would have to crawl over it. To help them inside was an astronaut, Franklin R. Chang; Sally for one was glad he was there, because the spacecraft was vertical instead of horizontal (which is what they were used to in the one-G trainer), and therefore a little disorienting. Still, it was familiar. McBride said that when he was strapped in his seat, he felt for a moment as if he were in a simulator. The illusion didn't last long, though. "This time, I knew it was real," he said. "The adrenalin was pumping. In sims, if things go wrong, we turn off the simulator and freeze the data. We can't do that here."

Just before the rockets ignited, Crippen told them all, "Get ready for the trip of a lifetime." No American astronaut sitting on the launch pad ever had a fresher memory of a launch, for his previous one had been less than six months before. The view out of the window was much better than the simulator graphics, which are highly stylized. The sense of lift, of tremendous power, of staccato vibration, was far beyond anything the motion-based simulator could do. "There was a lot of noise, a lot of vibration, and my heart doubtless was beating very quickly," Marc Garneau, the Canadian astronaut, said later, adding with a welcome candor, "During those moments I was a little bit afraid." Sally Ride compared the sensation of ascent to having what she called an "E ticket," as she had done during STS-7—only this time she said it was like having three E tickets. When I later asked her what she meant, she explained, "At Disneyland in California where I grew up, you could go on all the best rides if you had an E ticket, and we kids used to save and hoard for them." She said she had not been thinking of any particular ride at Disneyland: "I guess there's no ride quite like going into space," she said.

At main-engine cut-off, all the observers of the launch at Kennedy headed for home; the astronauts would make it to orbit. As he watched his friends ascend skyward atop a curly black column of smoke and listened to the crisp air-to-ground announcements ticking off the abort capabilities, Browder told me later, he was glad Crippen, and NASA, put so much emphasis during training on ascents. "It's only eight and a half minutes to main-engine cut-off, but those are the most critical moments," he said. "All that power! It's imperative that they have quick responses." McBride, too, thought of his training during ascent. "I kept waiting for all these awful things to happen, and they never did," he said. No engine got stuck in the bucket, there was no need for any of those aborts: return-to-launch-site, transatlantic, abort-to-orbit. There were some other differences from the ascent sims: as the spacecraft was heading east across the Atlantic into the morning sun, the glare meant that McBride and Crippen couldn't see much—a disadvantage the simulator had not anticipated. And the gravity forces of ascent (as much as three times that at sea level) put a greater weight on McBride's chest than he expected, making it hard at times to breathe. He had one malf to remind him of the sim world: after the first orbital maneuvering system burn, a yaw jet—one of four on each side of the vertical tail—wouldn't fire. On instructions from the ground, he switched it off permanently, together with its counterpart on the other side of the tail. There was plenty of redundancy.

In orbit, after they unstrapped themselves and began moving around, Dave noticed that he, Kathy, Jon, and the two payload specialists, none of whom had been in space before, were rather clumsy, bouncing off things. Crippen and Sally, on the other hand, "got graceful very fast." Crippen hadn't told them much during training about how to get about in space (he had given them a few tips, such as not to try to make swimming motions), and Dave felt that Crippen had been wise, for getting about in space is something everyone had to learn for themselves. "It's intuitively obvious," he told me later. "But on Day 1, you could easily tell the rookies from the old hands."

Everything went according to the crew activity plan for about the first six hours or so of the flight. In orbit, the astronauts went about their initial preparations like clockwork—stowing the five seats, far easier to do in space than in the one-G trainer, where they weighed ninety pounds, and opening the payload-bay doors. They powered up the orbital refueling system—it would make several hydrazine transfers internally, before the EVA. They aligned the inertial-measurement units. Over three periods of contact with the relay satellite, the Remote Payload Operations Control Center at Goddard checked out the earth-radiation budget satellite. During the third orbit, Sally Ride raised the satellite out of the cargo bay. The umbilical did not stick. The arm did not break. "The satellite is now close to the deploy attitude; in fact, it just got there," Sally told David Hilmers, the capsule communicator.

Hilmers, who could see it on television, said it looked very nice from Mission Control. "It looks very nice from here, too," Sally said gaily. It glistened, golden in the sunlight—if in the simulators it had looked like an automobile radiator, in space it was the radiator of some rajah's Rolls-Royce.

Before the actual deployment, the two payload specialists, Garneau and Scully-Power, were sent below decks to get them out of the way. Whenever anything that required concentration on the part of the crew was going on, such as the satellite deployment or an orbital refueling system transfer, the payload specialists were sent below. They didn't mind, or at least they said they didn't; Scully-Power told me later that during the integrated sims he and Garneau had learned what were the critical times for them to be out of the way. "Because of the integrated sims, everybody knew what he or she would be doing, and when. So the probabilities of conflict were virtually eliminated." Scully-Power felt that the integrated sims were so important for the payload specialists that in the future their training should always include the two-month period when a crew was doing them. Crippen encouraged them to help with meal preparation and with such maintenance tasks as changing the lithium-hydroxide canisters that purified the cabin atmosphere, but with very little else; most of the time they were on their own to carry out their own experiments and observations.

With the payload specialists packed away below decks, Crippen was delighted with the way the crew worked together on deploying the satellite, just as a crystallized crew should: "Sally was on the arm, with Dave assisting her," Crippen told me later. "Jon was flying the orbiter, and Kathy was photographing. Me, I was sitting back and managing." This idyll did not last very long.

Hilmers told Sally to go ahead and deploy the solar panels, but when she tried to do this, she got no joy—as had happened during one sim I had seen. This time, however, there was no Bassham to take pity on her and make her joyful when she tried again.

Hilmers reported consternation at the payload control center at Goddard. A backup method, in which the ground could deploy the panels, was tried; it failed. The ground established that the latches that held the panels in had released; the panels were just not springing into place the way they were supposed to. The ground had another idea: Hilmers asked Sally to jolt the satellite by rolling the end effector, with the satellite attached, to the right and stopping suddenly. Sally tried that three times, but without success; it was hard, she said, to get it rotating very fast.

Before losing contact with the relay satellite, the ground asked the astronauts to orient the spacecraft so that the satellite would be

The entire crew in flight, looking like a crowd scene in a Muppet movie. Crippen pointed out that his record-breaking crew of seven equaled the entire original astronaut corps. Top row, left to right: Scully-Power, Crippen, Garneau. Bottom row, left to right: McBride, Ride, Sullivan, and Leestma.

Kathy with a pair of binoculars at one of the forward cockpit windows, getting ready to do some earth watching, the crew's favorite pastime. As a geologist and member of the team of scientists for the shuttle-imaging radar, her special interest was impact craters, but she also sometimes served as forward spotter for giant eddies and whirlpools in the ocean, a special interest of Paul Scully-Power.

In the middeck, Kathy floats over Crip and Jon.

Marc and Dave in the middeck. Which one is right side up? The only way you can tell is by the labels on the cabinets —but then the entire orbiter might be upside down in relation to the earth, a common way for it to fly.

Marc Garneau and Paul Scully-Power conduct one of the Canadian scientific experiments.

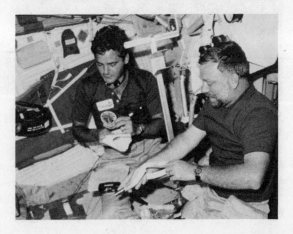

toward the sun, and asked Sally to make sure that one or the other of the hinges for the solar panels was always toward the sun. The thinking at Goddard was that the hinges had become so cold in space that they had contracted, squeezing the pins at their centers that they were supposed to swing freely around; they had frozen themselves shut. (The history of hinges on satellites is a constant to-ing and fro-ing between two principles: build the hinges small, and they break; build the hinges big, and they retain the cold so they won't open.) To carry out these instructions, Sally had to move the arm joints at a faster rate than the mission rules allowed—the permission to use the maximum rate came from Cox, who had himself conducted most of the analysis for the rules and knew they were very conservative. After one orbit of warming the hinges, Sally, back in contact with the ground once more, shook the satellite again, and the panels opened—first one, slowly, and then the other, more rapidly, instantly transforming the golden car radiator into a golden moth. (After the flight, to commemorate this success, Sally saved the manipulator-arm checklist as her souvenir of the mission.)

At simulations astronauts had practiced the satellite's solar panels not popping up. "We did such cases several times," Robert Bassham, the satellite instructor, said later. "On the occasion I'm thinking of, we didn't say the hinges were cold—we just said there was a mechanical failure, and that the primary and backup systems for springing the panels up didn't work. But the hinges could very well have been cold." (Precise causes of malfs, I knew, were often irrelevant in the sim world.) That time, Sally shook the arm, and it worked. When the panels popped in place during the mission, the chief of the remote control center reportedly said, "Hey, it worked just like the sim." Bassham said he gave everybody concerned an A: "I gave the astronauts an A, the satellite people an A, the flight controllers an A— and the instructors an A plus."

Although exact parallels between sims and actual missions do from time to time crop up, without a crystal ball sims obviously fall well short of predicting events on a real mission. Browder, Rainey, and the others drill the astronauts in as many areas as possible, in order to make them thoroughly familiar with the equipment and what might go wrong with it. The hope is that whatever does happen on a mission, they will develop the competence, perhaps even a sort of sixth sense, to cope with it. "And that in general is what you are going to find," Browder told me later. "You won't often see a malfunction identical to one on a sim, but you'll often see something close to it that gave the crew experience in that type of malfunction." There was much about the mission that was familiar from the sims; the mission was as close to the sims as many of the sims were to each other.

Cox decided to hold onto the satellite for one more orbit so that its controllers at Goddard could uplink (transmit from the ground) new data to it. Because of an error in their original loading of the software, the satellite's antenna would be aiming in the wrong direction after deployment; if they were unable to uplink new loads to activate their attitude-control system, they could lose the satellite. The Goddard controllers had been willing to take this risk, in part because they didn't want to throw the crew activity plan any farther out of whack and in part because they didn't want to cut any more into the imaging radar data takes. (One effect of the sims had been to make the different groups a little more considerate of each other.) However, the Goddard controllers were grateful to Cox for making the decision to delay. Finally, on the eighth orbit (two orbits or almost three hours late), Sally released the satellite with the gentlest of tip-off rates.

The crew activity plan was indeed out of whack, and it would get more so. Far more than a sim, a mission has a life of its own: you can't freeze it, and problems build on each other in a way that makes even the long sim look tame. To Bill Holmberg, the flight activity officer and keeper of the crew activity plan, this situation is one of the fundamental facts of life. Holmberg told me later that the plan—that paradigm of rigid prepackaging—worked perfectly for one and a half pages out of forty, up to the point where the solar panels didn't deploy, and it never got back on track again. When I asked him how he felt having a year's work disappear out the window at one stroke, he laughed. "In this business, a plan sometimes is good only until about when you lift off the ground," he said. "What's important isn't sticking to the plan, but building the document in the first place, for that process makes you think of the things that you want to do that might come up, and then you have the document there as a reference for everyone. It's a lot easier to rewrite something you already have than to write it from scratch." Holmberg felt about the plan very much the way Browder felt about his scripts for stand-alone sessions. NASA planners, flight controllers, instructors, and astronauts are extremely flexible, something they have learned from years of watching the wayward fashion in which hopes and plans turn (or don't turn) into reality. They have for a year acted in concert to bring about a space flight. In space flight,

NASA has frequently and perhaps unconsciously used human beings as a basis for new designs: the manipulator arm, which parallels the joints and bones of the human arm, is one example of this. Collectively, the people at Mission Control, in Building 4, and in the spacecraft may parallel the way a single mind thinks and acts—certainly, listening in on the loops is like tuning in on one's own thoughts. Together, as the flight controllers and the astronauts move from the sims to the real mission, they illustrate that the key ingredients in blending thought with action are continual projection of what might happen, anticipation of things that could go wrong, constant revision of a plan in the light of new information, and great flexibility—to the extent of scuttling all previous plans—when things fall apart.

Holmberg felt certain that the astronauts would get back on the timeline because he had put about fourteen hours of extra time in the first day, to make things more flexible. The astronauts were not scheduled to make their first orbital maneuvering system burn to lower orbit until the next morning, after the backup opportunity to deploy the satellite, and so the fact that they were three hours late getting the satellite off that evening was not serious. But then, as in any good sim, another problem arose: the astronauts deployed the Ku-band antenna (for transmitting scientific data) at the starboard side forward of the cargo bay, where the black dish swung outboard and aimed at the relay satellite; a computer would keep it properly oriented, adjusting the dish on its two axes so that by simple adjustments of two motors, one for the left and right, one for up and down, it could aim at any point in the sky. They also deployed the imaging radar antenna, which would be taking data in the higher orbit. They oriented the spacecraft so that it faced its target on the ground below, while the Ku-band antenna automatically continued to face the relay satellite. Data began flowing from the radar through the Ku-band antenna, to the relay satellite, to White Sands, and thence to the payloads control center in Houston. A stream of data rushed through for about two minutes, and then it stopped. The Ku-band antenna had lost its lock on the satellite. The motor for one of its two gimbals was completely dead; on one axis, the antenna was disengaged so that it bobbed up and down freely. The motor for the other axis worked, but sporadically. Something of the sort had happened during the long sim. Because the astronauts had had a long day and it was bedtime, Mission Control told them to knock off and go to sleep. The planning shift on the ground would wrestle with the problem overnight and give them the fix in the morning.

*T*he astronauts in fact never got back the use of the Ku-band antenna
motors, though in the course of the night the planning team at Mission
Control came up with a partial fix. The idea was to stop the antenna
from waving loosely about by unplugging a wire down in the middeck
that carried current outside to the antenna's motors. In order to reach
the wire, Sally and Dave had to move a row of lockers away from the
forward bulkhead in the middeck. The 41-G astronauts had not practiced this particular bit of inflight maintenance during their training.
"We never did anything physical in the sims that would make the astronauts move the cabinets," Jim Freehling, the communications instructor, said to me when I saw him later. "If you do that kind of comm
malf, of course, the data flow stops—and so does the rest of the training. It does tie your hands behind your back. So the whole business of
repairing the Ku-band antenna had to be pulled out of people's heads."
This was not entirely the case, for Sally had practiced something like
it when she trained for the STS-7 mission; hence, she took the lead in
the job now. Crippen remained at the rear windows of the upper deck,
and when the wayward antenna had swung around to what he thought
was the right position—at right angles to the spacecraft—he yelled
down to Sally to pull the plug. They stopped the antenna within one
and a half degrees of where they wanted it. Once it was still and the
flight controllers knew precisely at what angle it had stopped, they
could compute at precisely what attitude the orbiter should be oriented to lock the dish onto the relay satellite; that attitude, or course,
would change slowly as the spacecraft orbited the earth.

On the second day, after they had stopped the Ku-band antenna,
and after they had lunch, Kathy set about folding the radar antenna,
which had to be done before they could fire the maneuvering system
engines and descend from their original orbit, one hundred ninety-
eight miles high, to the intermediate orbit of one hundred forty-eight
miles, which was better for the radar and the other earth-watching
instruments. According to a mission rule, before the maneuvering
system engines could be fired, the antenna, a delicate and wobbly af-
fair, had to be stowed, lest the force of the burn set up oscillations that
would break it: since it was on one side of the cargo bay, the thrust
would pass through it at an angle. When it was open, the antenna—a
central panel and two side panels—was a blank white screen thirty-
five feet long and seven feet across that largely filled one side of the
cargo bay. It looked like a big, empty billboard. To close it, Kathy first

had to press switches that made the two side panels fold onto the center panel, and then she pressed another switch that caused the antenna to fold down into a box. A motor drove it down so that prongs on the central panel slid into sockets on the box; a microswitch would signal that the whole apparatus was securely closed. The signal never came. Kathy looked out the window and saw that the panel had not fully shut, which was tantamount to its not being shut at all. (During the long sim, Ron Weitenhagen had wanted to insert a similar malf, but it hadn't gotten in.) On advice from the ground, she used the backup system, with its own backup motor, but that didn't work, either. The third option was to fire the pyros, releasing springs that would slam the panel shut—but in that event the radar antenna couldn't be used again, and it was far too early in the mission to consider doing that. A fourth option was to disobey the mission rules and leave the antenna open and do the maneuvering system burn anyway, on the chance that the thrust wouldn't cause damaging oscillations. Kathy did not like this idea; when she had popped the antenna open the day before, it had wriggled and writhed, she told me later, "like the Tacoma-Narrows Bridge"—a classic case of dynamic instability (resulting from a high wind, not from an off-center rocket thrust) that tore the bridge to pieces.

There was, in fact, another option that saved the situation. On instructions from the ground, Sally Ride maneuvered the arm so that it was over the folded but unlatched antenna and then pushed down, closing the latches. "The ice busters strike again," she said, referring to the 41-D astronauts' unorthodox use of the arm to knock the icicle from the vent. The ground controllers were not entirely comfortable with this solution, because the arm had no feedback that told how hard it was pushing; consequently, it would be easy to strain one of its motors, or dent the antenna, without knowing it. Neither of these things happened. Browder told me afterward that, on the day before the astronauts left Houston for the Cape, they had simulated the problem of the radar antenna's not latching and had used the arm to press it down, just as they were doing now. That night, Kathy again tried to shut the antenna, unsuccessfully; the next morning, Sally again held it down with the arm to latch it. The maneuvering system engines were then fired once more to bring the spacecraft's orbit down to one hundred twenty-one miles, where it would remain for the rest of the mission. When Kathy reopened the antenna, she had another problem reminiscent of the long sim—the failure of the microswitch that held the antenna open at its widest extent, sixty-one degrees. They had to be content with fifty-seven degrees, using the orbiter itself to aim at anything further.

By far the greatest problem with gathering the radar data, however, was the failure of the Ku-band antenna on the evening of the first day, requiring that it be rigidly fixed in one position so that the orbiter could either collect radar data or transmit it to the ground but could not do both at once to transmit data to earth live, in a steady stream, as planned. (The other instruments that had to be aimed at the earth were similarly out of luck.) The situation, once again, was highly reminiscent of the long sim: if the instructors did not have crystal balls, I was beginning to wonder what is was they used instead. That episode in the sim, however, had only lasted a few hours, so that it had provided just a taste of what was happening now. As had happened in the sim, the radar data now would have to be recorded on tape; then the tape would have to be dumped (played back) to the ground. And the playback could not be sped up: it took as long as the time spent recording. There were seven tapes available, each one able to record twenty minutes of data. It was the situation with the Ku-band antenna more than any other that threw the crew activity plan out the window, for the well-laid list of attitude changes (one hundred two of them) that would enable the radar antenna to shift from one side to the other of the orbiter's ground track, in order to take in a variety of targets, was largely useless. The forty-four scientific teams and the engineers in the payloads control center, across from the Mission Operations Control Room, had to do some massive replanning for the entire rest of the mission. Wherever possible, they attempted to dump tapes over oceans so that they could take data over land. As things worked out, there were only about three twenty-minute periods every day when they could dump data, and consequently they only gathered one hour's worth of data every day. They attempted to give priority to those ground targets where scientists were actually present, getting what is called "ground truth" to compare with the data in space. Forty hours of data had been expected from the mission; only eight hours— 20 percent—was finally acquired. Much later, I asked Dr. Charles Elachi, the chief scientist for the radar, whether the replanning had caused much turmoil in the payloads control center—something I had been led to expect from some of the sims. The replanning during the mission had gone remarkably smoothly. "Tempers flare much more in the sims than in real time," Elachi told me. "In sims, you learn that everyone is under pressure, and that you need everyone's cooperation if you're going to get anything yourself. It took a long time for people to understand that. If we hadn't had the sims, we'd have been in serious trouble. It was there that we developed lines of communication between ourselves—who should talk to whom. If we hadn't ironed out our differences in the sims, we might well have wound up with no data at all." The sims indeed had worked a strange and marvelous alchemy. Holmberg said almost exactly the same thing: "The sims are absolutely essential because they teach us how to react; they teach the people on the consoles how to talk to their back rooms; they teach the people in the back rooms—such as ourselves—how to talk to each other. The reason the space program works so well is that we practice." The replanning of the radar data takes went so well that almost every scientist got some data; although they received only 20 percent of the data they had expected, they met 40 percent of their scientific objectives.

Although Kathy Sullivan was attached to the geology team that was gathering data on impact craters, she was too busy to spend much time looking for them on the ground, or even to notice when the radar was taking data and when it wasn't. (It was turned on and off from the ground.) She had to make sure that the tapes were changed on time, and every once in a while she would verify something on the ground for one or another of the radar scientists. For example, curious signals were bouncing back from the North Atlantic, causing an investigator to wonder if there were icebergs there. Kathy reported that there were. For another investigator she verified that snow covered much of Canada, and for a third that much of the Sahara was undergoing a sandstorm. Occasionally she sneaked a look at an impact crater (there are some big ones in Canada), but by and large she concentrated on other people's experiments, not her own. She in fact was doing exactly what made her happiest, and what had led her to apply to be an astronaut in the first place. If it was high seas and balky scientific equipment she craved, she was in her element.

Then on Monday, the fourth day of the mission, the relay satellite lost its own attitude control and hence its lock on the orbiter's Ku-band antenna, and what was worse, its lock on the ground station antenna at White Sands, so that it couldn't be controlled. It was freewheeling through space as if it had mutinied. All Ku-band transmission suddenly stopped. At the script session I had attended, a failure of the relay satellite, owing to its batteries running down, had been suggested, but it had been discarded on the practical grounds that the crew would need the satellite later. The present problem at first was attributed to a "cosmic failure—a solar flare perhaps, causing the satellite to lose its memory." An idea worthy of a sim-scripting session, perhaps, but one that did not prove any more durable. The satellite controllers at White Sands, who were completely at a loss, kept announcing that everything would be fixed in two hours' time. Two hours went by, then four, then most of the day. This was perhaps the low point of the mission. To Kathy, it was as if a wave had washed over her oceanographic vessel, carrying all the experiments away. The only people who didn't really seem to mind were the other astronauts; they were sorry for the loss of data, of course, but the failure didn't really affect them. "From my point of view, it didn't affect our activities as much as it did the ground's," Crippen told me later. "It had the ground in a tizzy, but not us. We just sat back and looked out the window." What had happened was that one of the satellite's sensors, which are meant to hold it in the proper attitude by always remaining fixed on the earth's horizon, had gone chasing off after the moon, which had risen right where it was looking. Whenever the moon rises, a technician is supposed to switch to another sensor, but he had not done so this time. It took sixteen hours to regain control of the satellite, during which all manner of data from the spacecraft was lost. Browder said to me later, almost paraphrasing what Freehling, the communications instructor, had said, "We never simulated a major relay satellite problem, or any other major communications problem, over a long period of time. Sim Sups don't like to do it because flight directors cry bloody murder—no data coming in, no training. But we have to review that reluctance. It's important to train the flight controllers, and the astronauts, for what to do when they have no data." I could see a fertile new area for future scenarios.

On Saturday—Day 2, the day after the problems with the radiation budget satellite deployment, the same day that the Ku-band antenna problem was being worked on, the same day that it was being discovered that the radar antenna wouldn't close, and two days before the relay satellite was lost—another problem appeared: ice formed in the vent for the flash evaporators, the main system that supplements the big radiator panels in dispelling heat from the cabin and electronics. The situation was reminiscent of the icicle on the 41-D mission, though in this case the ice did not involve the waste-management system. 41-G was about the first flight in which the waste management system worked perfectly. (The flight controllers could tell it was working well, because every time the astronauts used it, there was a drop in the cabin pressure.) But the waste management system was about all that was working. Aside from that one bright spot, the 41-G mission was turning out to have more anomalies, glitches, nits, and malfs than almost any previous shuttle mission. One instructor told me that the mission seemed to have about the same frequency of problems as a long sim—an exaggeration, but one that caught the flavor of things. And Browder said to me, when I saw him the day after the relay satellite broke down, "This whole mission has been like a long sim—whenever one problem is solved, you get another one." Ed Rainey told me that Browder had congratulated him on the great script he had written for the mission. John Cox told Rainey that the flight was almost a continuation of the sims: "When things go wrong on the mission, I sometimes find myself blaming Sim Sup," he told me around the middle of the flight. Rainey took all these comments as compliments; noting the number of situations that had been simmed and that later came up on the flight, he said to me, "It's a statement of the thoroughness with which the training folks do their job. If a problem on the mission is not exactly the same as the one we've simmed, it's usually close enough." Most of the problems that came up affected the payloads; only a very few were orbiter problems, and of these the most serious one was the icing of the flash evaporator. If it hadn't been resolved, the astronauts might have been forced to come home on the fourth or fifth day, Monday or Tuesday.

The flash evaporators, which supplement the radiators for dispelling heat, are simply coils of tubing where the freon, warm from the spacecraft, expands and releases its heat, very much as it does in an air conditioner or refrigerator. These are just inside the exit to space of a vent for dumping excess water generated by the fuel cells; as soon as the water hits the vacuum inside the vent, it evaporates instantly. The term is "flash evaporation"; if ordinary evaporation cools, then flash evaporation provides the ultimate freeze. It wrests the heat from the freon and the coils and rushes it off to space.

During the sims, I had seen various malfs and glitches thrown into the cooling system—Randy Barckholtz had sprung a leak in the water loop, and Keppler had hurled a meteor at the plumbing where the freon loops meet the radiators—but I had never seen anything involving the flash evaporators. They, indeed, had never before caused problems in space, but Keppler, the systems instructor, had anticipated that they might; during one integrated sim, he and Randy had thrown a malf into the flash evaporators—a mechanical problem, not an icing one. What happened now in flight was an example of how the sims can sometimes be too smart by half and can backfire. In the sim I had missed, Cox had recommended that the astronauts recycle the switch and do a restart—that is, turn it off and on again, just once, because that was all the procedures allowed. It had worked. Now, however, when Crippen recycled the switch once and it didn't work, he went on and recycled the switch six or seven times, without consulting the ground. He had got the idea from another simulation in which doing this had been successful, but that simulation had used a faulty computer load that gave him the wrong idea that he could restart the flash evaporators indefinitely without harm. This time, the vent became increasingly flooded with water, the way a flooded car engine does if you keep on restarting it. The water, instead of flashing, turned to ice—more and more of it.

The situation was serious, for if the ice couldn't be cleared, the astronauts would have to come home. Because the ice was inaccessible, it couldn't be knocked off with the arm. The malfunction procedure to try to clear the problem called for turning off all the freon—not only the loop that goes to the flash evaporators but also the ones that go to the radiators—in order to get the freon very hot and melt the ice. However, there is a danger that if the freon were to remain too hot for too long, it could rapidly overheat all the orbiter's electronic equipment, and a situation could arise in which the spacecraft had just forty-five minutes to be on the ground before the computers quit. Neither Crippen nor Cox liked that idea. "If you use that procedure, I hope you always have an emergency landing site right in front of the orbiter, in case we have to come back in a hurry," Crip said. They decided to look for other alternatives.

When Cox went off duty a little later, he went to his change-of-shift press briefing (during a mission, there is apt to be one twice or even three times a day), and when he dropped back to the Mission Operations Control Room, he found that the new flight director and the new flight controller for the electrical and environmental system had decided on their own to put the plan into operation, despite the fact that both Crippen and Cox were opposed to it. Each shift has to be able to respond to emergencies as it sees fit, and the plan was indeed in the procedures. Cox sat next to the electrical and environmental controller, and, as he put it later, "challenged" her to find another way; he urged her to see if, instead of using hot freon, she could use warm freon for a longer time.

And she did. By using the other cooling systems, she was able to keep the cabin just hot enough (without frying the astronauts) so that the freon for the evaporators would melt the ice over a couple of days. Accordingly, the cabin temperature was raised from the normal seventy-four degrees to about ninety degrees, and it remained there through the remainder of Saturday, all of Sunday, and into Monday. Crippen and the crew sweltered, though Sally said later that it was no worse than Houston in summer. After a couple of days of tropical conditions inside the spacecraft, the ice was gone. Crip was grateful; he most likely would have come home early rather than proceed with the more drastic remedy.

Browder, watching from the Simulation Control Area, where part of his job was to evaluate the sims in light of the mission, was enchanted; here was a problem and a fix the instructors had never thought of. It would be added to the instructor's bag of tricks, their repertory for future sims, along with the relay satellite that wandered off after the moon. Clearly, missions played into training as much as training played into planning and into missions; missions, training, and planning formed a tight, interlocking triangle.

When I saw Ted a little later, he was full of visions of future sims. "We're getting back to the chess game," he said. "Now that we know what the flight controllers' move will be when the flash evaporators ice up, when we sim it we'll try to be a step ahead; we'll slip in some reason why they cannot heat up the cabin. Maybe we'll fail a fuel cell and make them do an emergency powerdown, particularly of equipment that puts out the most heat into the freon loop. That way, we'll force a decision on the flight director of whether to power down essential equipment that otherwise might break, or whether to leave the equipment powered up for thawing the flash evaporators."

Browder felt that the astronauts' state of being was uniformly very good. The only time Browder detected a note of testiness in Crippen's voice was Sunday afternoon, when the cabin temperature had been about ninety degrees for the better part of a day and everything seemed to be breaking down. "I thought Crip was a little short," he said. "Yet there were seven guys in a small area, with the temperature at ninety—no wonder he wasn't in the best of spirits! He wasn't angry, but it was clear that all this was tiring him. In fact, I don't know why Crip wasn't madder! When my car breaks down during a long trip on a hot day, I get mad. The string of problems they've had has got to be frustrating. I've been paying close attention, not only to what Crip and the others are saying, but to how they say it—their inflections. Crip's very cool, not at all panicky, and I attribute the good behavior of the rest of the crew in part to him. The crew is doing very well—they don't seem to be hanging around watching developments in the latest problem, but rather they're all doing their duties, leaving the problems to the ground and to Crip, who avails himself of whatever crewman he needs." They had clearly not become uncrystallized in the heat.

After the mission, Crippen told Browder that what everyone had suspected earlier—that Crippen's absence in the first half of training resulted in a more versatile, cross-trained crew—had in fact proven correct and had paid off in space. (Both Crippen and Browder told me later that nothing came up during the mission to change their minds about it being all right for a commander to join a crew in the middle of training, provided that the commander was experienced and that there was nothing unusual about the mission requiring special training for him.) With respect to the greater flexibility of the crew, Crippen said to me afterward, "I took advantage of this ability not so much in fixing malfs, which Jon or I would handle, but in doing water dumps and other maintenance chores; I'd be happy having anyone who was free flick the switches. Or to reconfigure the oxygen-nitrogen system for our atmosphere. Or if a maneuver was coming up, I'd get a mission specialist to punch the numbers in the computer or select the correct computer program." There was no doubt that the crew's flexibility freed Crippen for troubleshooting. In this regard, both Crippen and Ted Browder thought very highly of all three mission specialists. It is possible that in the future, mission commanders won't necessarily always be pilot-type astronauts but mission specialists (NASA managers sometimes draw the parallel with ocean liners, whose captains are not the ones who steer the ship); and someone like Dave, Sally, or Kathy, all of whom are familiar with the spacecraft, could easily fill such a role.

Dave, in Ted's opinion, was having the time of his life in space. When I talked with Ted in the middle of the mission, he said he was having a harder time getting a fix on Sally, Kathy, and Jon, because they weren't talking all that much. Possibly they were feeling under the weather, something that happens for the first few days of a flight to about half of all astronauts. "It's hard to get a reading on Jon," Ted said to me on Tuesday, October 9, the fifth day of the mission. "Everything he says is businesslike. Yet he's normally very joyful. Perhaps he's too busy; he's had to help with most of the malfs. The new timelines and the new attitudes for the radar are his." When I saw Ted again a couple of days later, the Jon of the integrated sims was back on deck. "I'm most ecstatic about him!" he told me. "He really hit his stride. He appeared when I expected him to, and stayed out of areas where he shouldn't be. The pilot is never one of the most visible players, but Jon was always there as Crip's right-hand man—and that is just what a pilot is supposed to be." Later, Browder asked Jon if, during the flight, anything had come up that made him wish he had looked at it one more time in the simulators. Jon said there hadn't; he said that if there had been a serious failure in the orbital maneuvering system or the thruster system, or any other system he was responsible for, he had felt confident that he could handle it. Indeed, he had regarded it as a failure of the mission that there hadn't been more malfs, anomalies, and nits in his department, the electrical and propulsion systems, for him to solve—that the mission hadn't been *more* like a sim. Clearly, Jon's state of being during the mission was more than good; he was loaded for bear. Indeed, Cox did not need to consult Browder to know that the crew was doing very well. "Sometimes the crew is far ahead of Mission Control," Cox told me during the mission, applying a test Browder had used during the integrated sims. "We tell them to do a water dump, and they tell us they have already done it. We tell them to do a test of the flash evaporator system, and it's already finished. We tell them to get into the correct attitude for the next radar target, and they are already there."

Virtually nothing at all was heard from the two payload specialists, Scully-Power and Garneau, though they were in fact fitting in very well. Even though Crippen was reluctant to use them for tasks the other members normally did, and kept them out of the way during periods of high activity, he was pleased with the way Marc and Paul pitched in with the tasks they were invited to do, such as helping prepare meals or change the lithium-hydroxide canisters for purifying the atmosphere. Jon seemed to take them under his wing. Later Dave said to me, "We got along fine with Marc and Paul. They were great guys, but still, five plus two equals seven, and that's a crowd." He worried that since the flight had, in fact, worked out so well, NASA would decide that adding two payload specialists would be the way to go in the future.

The two payload specialists formed their own unit, part of but apart from the rest of the crew. They helped each other with their experiments and observations—in particular, Marc Garneau needed help, because the Canadian scientists had loaded him down with such a variety of experiments that it was lucky Paul Scully-Power was along to help him out. While Marc pointed, blindfolded, at targets on the wall—to test his ability in space to remember where things were and orient himself toward them—Paul noted how successful he was. Paul was also the subject of a variety of medical experiments that Marc performed on him, many having to do with the vestibular sense and its correlation with the eyes. The two spelled each other at the overhead window, usually oriented (along with the payload bay) down toward the earth. When it was dark, Paul was apt to be there, shooting pictures for Marc of auroras and the other glows in the earth's atmosphere, or the one on the orbiter's vertical tail.

In the daylight, whenever possible, Paul tried to be at the window, to look at the oceans—the main purpose of his being in space. With the orbiter upside down, flying nose forward, one of the best places for observing was one of the six forward windows that wrapped around the commander's and pilot's seats; upside down, these angled inward underneath Paul, who felt he was riding in the gondola of a blimp. The earth was not entirely cooperative; there was 70 percent cloud cover over most of it, as opposed to the 30 percent that had been expected (it was the same for both hemispheres), and Paul was disappointed that he was unable to see any of the major oceans clearly from end to end or side to side. The Mediterranean Sea was visible in its entirety, though, and as the spacecraft ran down it, he was able to see that the spiral eddies that marched from one end of it to the other were tightly interconnected, like a dozen gears in a row.

He observed spiral eddies along the edges of the Gulf Stream, too: perhaps the Gulf Stream spun them up, or perhaps they helped propel the Gulf Stream along. He saw solotons (the waves between horizontal underwater layers) in such places as the approaches to the Strait of Gibralter. "The advantage of having a man up there is that he can notice things that you otherwise would not aim a satellite at," Paul

told me after he got back. When he wanted to photograph some phe-
nomenon in the ocean—a spiral eddy, say—he would do so from the
overhead windows, which afforded a view straight down, though a
fairly restricted one. Because he couldn't always tell when a target
was coming, he would station Marc, or any other crew member who
happened to be available, up in what was called the "gondola," as forward spotter; he or she would give Paul advance warning. In this way,
the entire crew put in a lot of time at the windows—and it was a valuable sort of time, because it is often better when traveling to have a
purpose, something to look for, than to do random sightseeing.

The only people who were seriously dismayed by the relative silence of the payload specialists was a large contingent of Canadian
reporters who had arrived en masse to cover the flight of Canada's first
astronaut in space. Their editors had not sent them to Houston to write
about a lot of dead air. At the change of shift briefings, the Canadian
reporters badgered the flight directors—John Cox or his alternate,
Cleon Lacefield—to get Canada's first astronaut to say a few words.
The message finally went up, but very little came back in the way of
Canadian insights from space. One Canadian reporter referred to his
national hero in print as "The Right Stiff." During an inflight press
conference held on Tuesday, when the Canadian hero was visible for
about the first time, the Canadians went for him. Where had he been?
Had NASA placed him under wraps? Marc explained that NASA had
not muzzled him, but rather his experience in the Canadian navy led
him to believe there was no point in prattling on. He explained further,
at another press conference after he got back, that while he was aloft
he had almost forgotten that there was a contingent of reporters at
Houston with a special interest in him, as opposed to the others
aboard. He said he regarded himself now completely as a member of
the 41-G crew. The two months' integrated sims had clearly done more
than teach the two payload specialists to stay out of the way. Being in
space also seemed to be exerting a strange alchemy of its own.

*T*hough the entire crew generally did not chatter as much as they had
during many of the sims, there were many moments that were reminiscent of the carefree days of the stand-alones. The crew's occasional
television shows for the ground were often jolly times. They enjoyed
taking pictures that made themselves look even more crowded than
they really were. On occasion, they sat so close together that they

looked like lots of Muppets in a small room; one reporter remarked that they resembled a can of anchovies. The crowd motif provided Sally Ride with the idea for the crew's chief contribution to the repertory of funny movies from space, ideas for which had been running low: she set the Airflex movie camera up in the middeck, put it on automatic, and had all the astronauts follow one another down through the hatch: Crip, Jon, Sally, Dave, Kathy, Paul, Marc, Crip, Jon, Sally . . . After floating down through the hatch on one side of the middeck, each astronaut floated off-camera and up on the hatch on the other side and down again in an endless procession. "I defy you to count the crew," Sally said, and Crip remarked, "One way to have seven aboard is to keep them moving." It was the impression of some observers that they were propagandizing for smaller crews.

They took a lot of pictures, just as Stephen Seus had told them to. Sometimes they messed things up. They took numerous interior shots with the thirty-five millimeter camera, covering the flashbulb with a handkerchief, as Seus had taught them, to make the light less harsh; however, they forgot to open the lens a little wider, so many of these pictures were underexposed. And someone left the forty-thousand-dollar Airflex movie camera with its power switched on so that the battery drained; consequently, there were no good out-the-window films of Kathy's and Dave's EVA. Then they had used the Aero-Linhof, the wide-angle camera meant for earth photography, in a way that was less than wise: because of its wide field, Seus had drilled into the crew that the Aero-Linhof was only meant to be used through the overhead windows for taking pictures straight down toward the ground. Whenever the orbiter came within sight of California, through whatever windows, Sally, Dave, or Kathy—all of whom came from that state— grabbed the Aero-Linhof and fired away. Since they were almost never directly above California, all of these shots, about 60 percent of the total, were obliques, and they came out badly. There were also several obliques of West Virginia. "I wish they hadn't done that," Seus said to me later.

Sometimes when they made mistakes, however, it wasn't their fault. The videotape recorder operated differently from the one the astronauts had used in the simulators, so that they failed to record some television movies, and some data, properly. Later, Crippen declared stoutly this was not a crew problem: simulator equipment and flight equipment should be identical. This, of course, is the stuff that flaps in Building 4 and Building 5 are made of.

Dave and the levitating water drop: In weightlessness, water forms perfect spheres which make perfect lenses. The blur in the background is Dave's face. The camera is focused on the ball of water, where his face is reflected, upside down and distorted by the sphere, but with crystal clarity.

Yet, the astronauts got some superb photos of the ground, including some fine ones using the sun glinting off the ocean to bring out various features, such as a surge of water through the Strait of Gibraltar, which may have shown the influence of Paul Scully-Power. Jon McBride did an outstanding job with the IMAX camera; he was frequently to be seen attaching it to an overhead window, which it completely filled, to photograph the earth. The astronauts got so accustomed to the IMAX, and to Jon's lugging it around, that they treated it almost like a large dog; whenever anything interesting was going on, one or another of the crew members would say, "I hope Max is getting this." Max got the earth-radiation satellite just after it was deployed (at the moment Cox and Crippen had previously agreed upon), and Max got some footage of the EVA. Mostly, though, Max got the earth, as Graeme Ferguson, the head of the company that owned it, had asked—in particular, one marvelous scene coming down over England and France, and then down the length of Italy, all unusually clear, all immediately recognizable. Of this view, Jon told me later, he thought to himself, "Hey, this is it! This is what it's all about!"

Although in their other photography the 41-G crew did not rank as highly as some other crews, there was one series of thirty-five-millimeter pictures that Seus feels ranks with any pictures taken by any shuttle astronauts, in that it shows artistically something about space that had never before been demonstrated. Seus, in fact, had given them the idea during training, when he had shown them some photographs in advertisements of people standing behind crystal balls, which acted as lenses so that their faces were reflected, upside down, on the front sides. Water in space, of course, forms even more perfect solid spheres—and lenses. In space, Kathy, Dave, and Sally discovered that if they stood on one side of a sphere of water, the same thing happened. They reached for the thirty-five-millimeter camera. "There's this wonderful series with each of them standing, a little out of focus, behind a water sphere," Seus told me. "But the bubble is exposed just right, with Kathy's, Dave's and Sally's upside-down faces in miniature peering out at you, crystal clear."

The EVA was put off from the fifth day of the mission until the seventh day, Thursday, October 11, so that while they were out Dave and Kathy could do a repair job on the Ku-band antenna that would allow it to be stowed from inside the spacecraft just before entry. If they were unable to make the repair, so that the payload-bay doors could not be closed, Kathy and Dave would have to stow the antenna during the EVA; consequently, the later they went out the better. Similarly, if the flight controllers couldn't assure themselves that the radar antenna could be closed for the return to earth from inside the spacecraft, then the astronauts would have to close *it* permanently, too—and the later they did that, the better. As it happened, another mission rule required that the radar antenna be stowed while the astronauts were on EVA: in the first place, the open antenna was so delicate that a clumsy astronaut could damage it; in the second place, it couldn't be sensing the earth while the astronauts were outside lest they be radiated. (Dave, with his sharp nose for potential problems, had expressed to me his concern that he might be zapped.) Hence, before the EVA started, the astronauts and the ground had the opportunity to see if they could close the radar antenna once more from inside, using a new method, which, if it worked, would make everyone feel more comfortable about opening it again once the EVA was over. Because the prime motor and the backup motor were each too weak to shut the antenna by themselves, the engineers in the radar control room had decided to try using them together. The plan, which had never before been simulated, worked. Kathy said to me later, "We don't sim for things like 'The motors are undersized.' We would regard that as unrealistic, because we assume the design is correct. And in this case, the motors were designed to do the job gracefully and smoothly, which is usually the right way to go in space—not to use brute force." Cox thought the radar antenna had been overdesigned, with too many backup systems and microswitches to guard against astronaut error; if the designers had had confidence in the astronauts and built it sensibly with big enough motors in the first place, the problem never would have arisen.

Kathy and Dave had begun their prebreathing the day before, using their launch-and-entry helmets to supply pure oxygen. Hence, when they opened the hatch, snapped their tethers to the ones on the slide wires, and went outside, they were perhaps as bubble-free as any astronauts in history—though no one who had listened to their effervescent, spirited conversation would have had the slightest inkling that their blood was as flat as bottles of ginger ale whose caps had been left off for many hours. Dave came out the hatch first—with his round glass faceplate, and little lights at this ears, and the sectioned body of his inflated spacesuit, he looked like a fat white glowworm emerging from a hole. "It's like the difference between sitting at a desk in a big room and sitting at a desk in the middle of the prairie—you can see so much more," Dave told me later. "You get a real rush from that."

Getting used to the suit in space took a little doing (Dave told me later that it took him about half an hour to get thoroughly accustomed to it), and since most astronauts have difficulty adapting, their timeline gives them easy tasks for about the first half an hour so they can adjust. Although most astronauts have high praise for the accuracy of the WET-F, with respect to the neutral buoyancy they can attain there by means of carefully adjusted weights, there are of course ways in which the WET-F is quite different from space—or, as Kathy once said to me, ways in which "the WET-F lies to us." Nothing drops to the floor of the payload bay, as it does in the WET-F, unless it is shoved in that direction. In the WET-F, it had taken Kathy and Dave several tugs on the hand rails to get from one end of the payload bay to the other, because of the resistance of the water; in space, it took one pull. "You coast until you stop yourself against the opposite bulkhead," Kathy said to me later. "In time, you learn to adjust your movements. You learn not to move very fast or to push off very hard, for whatever motion you impart to yourself, you have to take out later. So moving slowly is the way to go." The learning process was impeded by a physical variation: in space, the difference between the pressure inside the astronauts' suits and the pressure outside was greater, and this made the suits stiffer. Arms, legs, and fingers were harder to bend. Kathy may have had more difficulty than Dave, for she said to me later, "Once we got out there, we could see much more, and looking at the earth from there was nice; still, we were in a machine, and the machine took a lot of our attention." Most of their tools were in a cabinet set into the floor of the payload bay, and Dave hovered there, removing equipment, like a bottom fish scavenging a river bed. Kathy and Dave mounted the small frames on the fronts of their spacesuits to which they attached the tools or tool caddies. They got out the special tools for the orbital refueling system. And then they whisked down the slide wires to the aft of the payload bay.

After Dave had stepped into the foot restraints in front of the refueling system, he made a dismaying discovery—the flashlight he had invented wouldn't light up. He had noticed this inside the cabin the day before, but he had replaced the battery and unscrewed the bulb and screwed it in again, then had recycled the switch several times in the best Cox-approved manner until he had got it working; now it had stopped again, and there was nothing he could do about it.

He felt as if he were back in the WET-F—indeed, as had also happened in the WET-F, the foot restraint came loose. (Unlike in the WET-F, he said later, the rest of the tools worked perfectly.) As it turned out, he didn't need the flashlight, though he was glad to use its mirror; in space, the sunlight reflecting into the refueling system illuminated its interior much better than had been the case under the chlorine-green water. On the dark side of the earth, Sally Ride positioned the manipulator arm over Dave's shoulder so that a light on its wrist (for the television camera there) gave him all the light he needed. The television camera gave the astronauts inside the cabin, and the flight controllers in Houston, a better view of exactly what Kathy and Dave were doing inside the refueling system than had ever been available during training. They had not used the manipulator arm or its camera in the WET-F. The arm, long and sinuous and with the light and lens near its end effector, looked like an inquisitive interplanetary bronotosaurus that had sneaked up behind the astronauts.

They chatted along as if they were in the WET-F. "Aha! Look at that! I've got it!" Dave said, holding up a piece of wire in the needle-nosed pliers. He worried that there might still be more wire down there, but a check with the mirror showed that it was all out. One problem that hadn't bothered him in the WET-F, but was a perpetual nuisance in space, was that a pair of brackets on hinges kept slamming back and forth in front of the pipe he was working on—they were braces that had been secured in front of the refueling system before launch, and would be fastened again by the astronauts for the landing. Dave told me later that they hadn't caused any trouble in the WET-F because they didn't keep slamming back and forth in the water where their motion was damped out (in both senses of the word), and hence no one had thought to provide a way of fastening them open. Dave kept batting them out of the way, which only made them slam back and forth with greater speed. "Perpetual motion," he said crossly.

Jon McBride told him to take a break—Dave was overheating a little. "You're right over a beautiful part of Canada—no, Cape Cod! Look! There's Long Island, New York." Astronauts often get confused about what particular geographic feature they are passing over; when Dave later narrated some of the IMAX film clips, he had to correct himself several times.

They were indeed flying over the Boston–New York corridor. "A lot of Sullivans down there," Kathy remarked.

He twisted the ball valve on (it was while he was doing this that the foot restraint broke off), then he stuck the next tool down through it. This time, it fit. He twisted it on, spinning the handle easily with one hand. "When did you ever do that in the WET-F?" Kathy asked. Twisting any kind of threaded tool or bolt into place was much easier than it had been on earth, for there was no weight on the threads. Conversely, when he had to turn valves, opening or closing them, the effort was much greater than it had been on earth, where the pipes had never been pressurized; here, with the hydrazine inside, the pipes were pressurized at times to five hundred pounds per square inch. Jon told him to knock off from time to time to take a rest, and on those occasions Dave would lean back in his foot restraints and look at the earth, as he had often pretended to do in the WET-F. By and large his simulated descriptions of the imaginary earth in the WET-F were better than his descriptions of the real thing from space.

"Every time you lean back like that and look at the earth, we get a great view of your shoulders," Sally grumbled from the window.

"No more earth watching!" Jon told him. "Get back to work."

Every now and then Dave leaned back so that Kathy could snap a picture to document the progress of the operation, and he would sneak more looks at the earth. Because of the need for documentation, the work with the refueling system took an hour and a half, whereas the WET-F, where Kathy had had no camera but instead had just held her hands up to her face and said, "click, click!" it had taken only an hour.

At one point, Dave looked back at the rear cockpit windows and noticed that one of them was empty—rare during an EVA. "I hope they aren't eating our lunch," he said.

"You'd have loved it," said Sally, down in the middeck. Being weightless did not affect her one-liners.

When they had finished the refueling test, they turned their attention to the Ku-band antenna. Kathy took the lead in repairing it: from the start, she had been assigned the role of contingency lead. When John Cox had asked Jim McBride (the WET-F instructor who, during the EVA, was sitting with Browder in the Simulation Control Area) whether he felt Kathy could handle it, he said she surely could. They would have to move the antenna on each axis by hand so that a pin, activated by a switch at the rear console inside the orbiter, shot into place, locking the antenna securely for the landing. If they could do this, they would leave the antenna swung in its outboard position so it could continue to transmit data to the relay satellite; the motor for swinging the entire antenna inboard so the payload-bay doors could close was in good shape. If they couldn't get the locking pins in place, the antenna would be swung inboard, deactivated, and secured for the trip home. Cox had spent much of his off-shift time working on the problem; he and an astronaut who had spent a lot of time working on EVA matters, Major Jerry L. Ross, had examined a duplicate antenna and discussed fixes. Cox worked with the computer graphics technicians to try to find the best position to secure the antenna for the trip home. (Cox told me later that 41-G probably required more fixes developed on the ground than any other shuttle mission; much of this work involved simulations.) Finally, the day before the EVA, the repair job had been simulated in the WET-F by Jerry Ross and George Nelson, the jet-propelled repairman on the 41-C mission who disparaged tethered EVAs; Nelson was serving as capsule communicator during the EVA. Sally Ride, of course, had helped out Nelson when she had simulated capturing the Solar Max satellite with the arm—the assistance that astronauts from different missions lend each other, in simulations or in passing on experience through conversations, is endless and interlocking. The day before, when Nelson had been briefing Kathy and Dave about what they had to do, and when they had been confused by a complicated diagram, Nelson broke off his explanation and said, "When you're out there, just go look at it for five minutes, and you'll see what has to be done." Dave said to me later, "That's the best piece of advice I ever got." He added that NASA managers generally don't like astronauts to do things so naturally because "they don't like to admit we can do things we haven't trained for." Cox, however, told me that astronaut crews are sometimes "notorious for refusing to agree to even *try* to do anything inflight that they are not trained for, which is why the ground goes to such lengths to provide them with as much helpful information as possible." However much astronaut training may encourage astronauts to think and act on their own— the aim of any good educational system—an organization such as NASA, which attempts to think out everything in advance and succeeds in this to a remarkable degree, must feel threatened when astronauts do exactly that. At the same time, no complex mission could be carried out without the fullest involvement of both the people in space and the people on the ground.

The job took a lot of coordination between the astronauts inside the *Challenger* and the astronauts outside. The previous Saturday, when Sally and Dave had moved the cabinets and unplugged the electric wire that ran the antenna's motors, they had also unplugged the mechanism that drove the pins to lock the two axes. Inside, earlier that morning, Sally and Jon had moved the same cabinets and had rigged a jump wire that would allow them to reconnect the power to the pins but not to the motors. The jump wire came in two sections that had to be plugged together in the middle to complete the circuit. (When Sally and Jon had first got it set up, they looked at each other in horror because both plugs were female; they hastily remedied the situation by rigging an adapter pin.) Outside, when Dave and Kathy looked at the Ku-band antenna, they indeed saw what had to be done. Dave reached over the side and rotated the dish, first on one axis, and then on the other. Kathy, who was watching from the port side, called out to Crippen when the pins were lined up with the holes they were meant to slide into; Crip shouted down to Sally in the middeck and told her when to plug in the two ends of the jumper. Sally worked the current in pulses, plugging and unplugging the cord, so that the pins were in effect hammered into place—the ground thought that was the surest way to drive them home.

That task done, Kathy took a look at the folded-down radar antenna to see if she could figure out why it was so difficult to close. It looked like an overstuffed sandwich, the filling being insulation that on the ground normally lay flat but in space seemed to have billowed, making it fatter than it should have been. "The insulation is billowing enough to interfere with a single-motor closing," she reported "And you don't need to miss by much to keep the latch from shutting."

After the EVA was over, everyone was tired. An EVA is exhausting for everyone—not only for the astronauts outside, but for the crew members inside, who have to concentrate intently on what is going on, sometimes gazing out the window for long periods into the bright sun. Even Crippen was exhausted. When the other astronauts went to bed, he was still only eating his lunch—a rare departure from his strict adherence to proper meals eaten together. During the day, he had actually missed making some attitude maneuvers, something that in the simulator he never had to be prompted to do. And, during the EVA, one of the screens for the computers had broken. When he had discovered it, Crippen had said, "Shannon must be on board." Someone else said, "She must have brought her light pen with her." This was the only time the astronauts while in space overtly crossed the line from the real world back into the sim world. It was the sort of malf Shannon had given them over and over again, and the astronauts had told her repeatedly the nit was unrealistic, that it could never happen

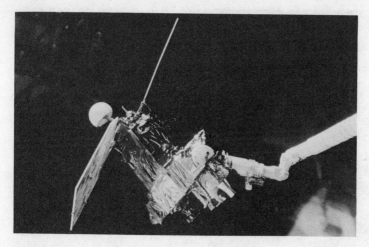

The arm is about to deploy the earth-radiation budget satellite, whose solar panels have finally popped into place. Wrist cocked, the arm might be about to flick a paper plane into the void. Sally Ride released the satellite with the gentlest of tip-off rates —that is, with hardly any wobble.

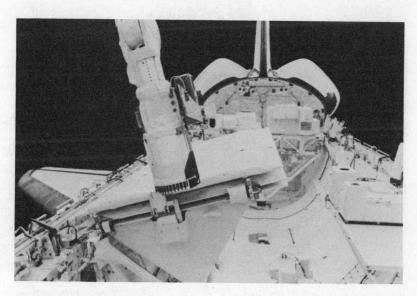

The ice-busters strike again: The arm—which on the previous mission, 41-D, had been used to whack an icicle off a vent—on 41-G holds down the antenna of the shuttle-imaging radar, so that it could be latched.

Kathy and Dave are at the aft end of the cargo bay, working on the orbital refueling system. The wrist camera near the end of the arm records the delicate operation, as if the arm were an inquisitive interplanetary brontosaurus peering over their shoulders.

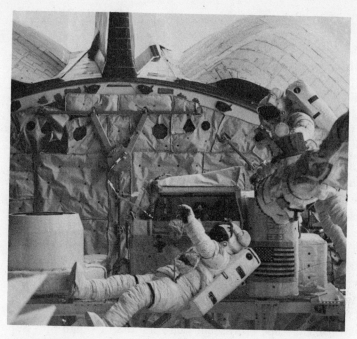

Dave and Kathy, and the inquisitive arm with its wrist camera, at the orbital refueling system. Dave appears to have slipped out of the foot restraints while twisting one of the tools into place.

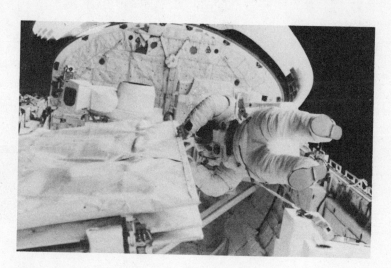

Kathy inspects the antenna for the shuttle-imaging radar, to see why it was so hard to close.

in space. At the end of the day, the screen was still down. Crippen listened while the ground read up a suggested procedure for fixing it. The ground was surprised; normally Crippen would have repaired a computer problem like this before the capsule communicator could read the fix to him. Not only did Crippen listen patiently to the procedure, but he didn't try to repair the screen that night. He said, "No, thanks! Mañana!" Putting off a repair until the next day was even more un-Crippenlike, and was clearly a measure of everyone's exhaustion.

The next day, their last full one in space, they were rewarded with a telephone call from President Reagan, but it might as well have been from Jack James. The main difference was that, instead of calling from an American Legion hall in Des Moines, he was calling from a railroad train outside Dayton, where he was campaigning—his schedule, not the astronauts', had slipped. There was a constant clickety-click and an occasional screech of a whistle in the background—they seemed like sound effects from an old radio program and were certainly within the scope of Mission Control's abilities, had Sim Sup known about the train in time for the sim. Reagan asked Kathy whether her space walk had lived up to expectations; told Sally she was getting to be "quite a veteran in space"; and wished them all a safe landing the next day. They all wished him a safe trip, too.

Smiling and Nodding

*T*he astronauts landed at the Kennedy Space Center on Saturday, October 13, at 12:26 p.m. The de-orbit burn and the entire entry went without a hitch. The yaw thruster in the tail, which (together with its companion on the opposite side) had been turned off early in the mission, caused a minor flutter in the propulsion controllers' back room: of all the jets used for entry, the ones used for yaw are the most important because they have to be employed farther down in the atmosphere than the others. (Because the spacecraft is tipped up at a forty-degree angle during most of the entry, the rudder at the back of the tail, which takes over yaw control in the atmosphere, is out of the air flow and cannot be used until the orbiter is flying more horizontally. This doesn't happen until it is almost in sight of the landing field and flying slower than hypersonic speeds.) While the propulsion back room was keeping an eye on the yaw thrusters, Jon had time to admire the view as the craft sliced across Canada, over Chicago, and thence south and east toward Florida. "I'm glad I got a chance to show Crip West Virginia," West Virginia's Ambassador of Good Will said later. Kathy, who was in the middeck, next to the hatch, took a series of pictures with the thirty-five-millimeter camera during entry out a round window at her side. (The only people in the spacecraft who during ascent or entry never got a window seat were the two payload specialists.) The series began over Canada and showed a variety of Canadian and American landscapes, some cities, some clouds, the Atlantic Ocean, Cape Canaveral, and finally a group of palm trees as the orbiter came to rest on the runway. The pictures, which Seus ranked with the water sphere pictures as unique in the annals of astronaut photography, were like a series of photographs taken through a porthole of a ship sailing into some tropical harbor.

I watched the landing on television in the conference room on the second floor of Building 4, where some of the instructors had gathered: it was the same room where the scripting meeting for the long sim had been held. Ted Browder wasn't there—he was over in the simulators, standing by in case he was needed to practice any last-minute changes in the wind profiles for the landing; he had spent much of the day before rehearsing them with John Young. The winds turned out to be fine all the way down—there wasn't a trace of a crosswind at the runway. I sat next to Andy Foster, the propulsion instructor on Ted's team, who hoped that the people at Kennedy would check the runway for alligators (an old wheeze among space hands). I asked Andy if he had ever simulated an entry without a yaw jet or two, and he said he had. "I've hammered them so hard about missing yaw jets at entry that I'm not worried about them now," he said. We were all waiting for the astronauts to come out of the blackout period, which they finally did over the border between Canada and the United States. Crippen told Mission Control, who had kept putting in calls to the spacecraft, that he could hear them "loud and clear." As each stage of entry was completed successfully, Andy said, "Super! All *right!*" When the orbiter had descended to one hundred thousand feet and was one hundred nine miles from the runway, it came in sight of the cameras at the Cape so that it appeared as a shining blip on the screen. "There she is! All *right!* Super!" Andy said. As it got bigger, puffs of smoke left white dots on either side of the spacecraft: the remaining yaw jets on the vertical tail were firing. The spacecraft, appearing now like a big, shining silver arrowhead on the screen, began sweeping around on its final turn, which would end with the shuttle lined up with the runway. Andy, who was helping it with body English, said, "Here she comes! I can see her speed brakes out. Super!" The speed brakes didn't jam, and a fire engine did not flatten a TACAN shack. Andy said, "Come on, Crip! Come on! Nose down! Nice! Real nice!" Andy's body English helped the orbiter through its final flaring maneuver. There were cheers and rebel yells from the crowd at the Cape as the wheels touched down. Andy gave a final "All *right!*"

Keith Todd, the arm man, came over and congratulated Andy on the nice landing. "The orbital maneuvering system and the thruster system went fine," Todd said. Andy congratulated Todd on the good arm work in orbit. Vicariously, they were still living the mission to the hilt.

Dave Hilmers, the capsule communicator, was congratulating Crippen on finally making it to Kennedy, on his third attempt. "You landed at Kennedy, but the beer is at Edwards," he said.

The beer, in fact, was at neither of those places, but at Ellington, with Browder and the team.

The beer almost didn't make it to the astronauts at Ellington, either. It was a stormy evening, so the welcoming ceremony was inside a hangar; about two hundred people were there—the astronauts' families; the flight controllers, instructors, and technicians who had helped with the flight, and *their* families; and management people. When Browder and the team got there at about six o'clock and tried to get into the V.I.P. section, a guard threw them out. Perhaps he thought they were underdressed, because the four of them were wearing the 41-G T-shirts they had bought at the Cape, which were a little tacky. Or perhaps the guard didn't like the looks of all the beer they were carrying, in coolers. Ted was hurt. He said, "We *are* V.I.P.s. We trained them!" That cut no ice. Then Ted said, "They told us to bring this beer!" The guard didn't relent until Glynn Lunney, then manager of the shuttle program and a certifiable V.I.P., recognized them and interceded on their behalf.

The crew was so big that it took two planes to bring them back to Ellington. The first plane, which taxied into the hangar at 6:30, carried Crippen, Sally, Kathy, and Dave. As they got off the plane, Ted handed each a beer. As the astronauts greeted their families, the team retreated to a discreet distance—but not so discreet as to be out of view. At length, Sally spotted them and came running over. Then Crippen saw them and followed. Soon all four astronauts and all four team members were talking excitedly. "They were like kids bubbling to tell us what they had done," Ted said. "They said that their flight was something they couldn't put into words—that you almost had to experience it to understand it, that they almost had to talk to someone who had done it before, to communicate fully." If the astronauts had had an experience that finally separated them from the people who had trained them, their manner in no way showed it. "They were so excited, it was hard to talk details," Ted said afterward. "They were extremely tired. But they said they couldn't wait to do it again."

The second plane, with Jon at the controls and carrying Paul and Marc as well, arrived half an hour later, because of the storm. The welcoming ceremony had to be repeated all over again; this time, Ted had the cooler full of beer knocked from his hands by the Canadian press, who stampeded all over him to interview their national hero. None-theless, Ted managed to throw Marc a beer while he was answering a thousand questions. Ted felt that Marc was in a tough position, being the Canadian man-of-the-hour on the one hand, and a 41-G crew member on the other. The rest of the crew helped Marc out by coming over, one by one, to shake his hand. "Another crew would have just fed him to the lions, but this crew was different," Ted said. "They now felt Marc was one of them, and they supported him." Ted clearly felt the same way: he wasn't handing out beers to just anybody this evening.

Ted and the team didn't see the astronauts again until a few days later, when Crippen and the crew gave a party at the Clear Lake Country Club, a mile or so from the Space Center, for the people closest to the mission. At the party Crippen presented the rookie astronauts who were rookies no longer their gold pins, similar to the one on the 41-G emblem, which are awarded only to NASA astronauts who have flown in space. Marc and Paul were not eligible, but Jon, Dave, and Kathy each got one. Kathy broke down when Crippen gave her hers.

Browder and Crippen talked a little shop at the party. Browder wanted to know if anything had come up during the mission that hadn't been covered in training. Crippen said that inevitably there might have been a couple of minute details that nobody had thought about ahead of time. Then he rephrased Ted's question: "If you ask, 'Is there any area where we were not trained to cope,' the answer is a resounding 'No!'" Crippen went on to say that even if something didn't come up directly in training, they at least had talked about it afterward, during the debriefings or even in the hallways. Naturally, there were a lot of things they had trained for that fortunately hadn't gone wrong on the mission—the failure of an orbital maneuvering system engine requiring a single-engine de-orbit, the broken computers requiring restringing, the fuel cells that overheated, requiring an emergency powerdown. Obviously a crew had to know these things, and they would probably happen on other missions; indeed, some already had. The comprehensiveness of the training, of course, explains why (even with no one's having had a crystal ball) the mission had so uncannily echoed the sims: almost everything that could conceivably go wrong had been thought of to one extent or another, and hence it would have been difficult (though of course not impossible) for the real mission to go very far outside the scope of the sims. Crippen confirmed to Browder that if anything *had* come up that hadn't been anticipated or talked about, their training would indeed have given them a general familiarity with the orbiter, a flexibility, and an instinct that would have seen them through. Crippen and the rest of the crew gave Browder and his team a high rating, especially their particular blend of sticking to a well-thought-out script and scrapping it to improvise: it encouraged a combination of disciplined knowledge and flexibility that is invaluable in space flight. Crippen told me later, "I thought they had just the right balance."

About a month later, the crew had largely ceased to be a crew. The payload specialists had already left. The others were going off in different directions on their postflight activities—the chicken dinner circuit. Jon was about to leave on a publicity tour of West Virginia. Crippen had moved out of the crew's office on the third floor of Building 4 and into an office of his own on the eighth floor of Building 1, the main administration building; he had been appointed deputy director of Flight Crew Operations, assisting George Abbey, though he was still a member of the astronaut corps. "The crew breaks up, and it probably should," Crippen said when I visited him in his new office. All the astronauts (Crippen, too) were awaiting new flight assignments. The first to get one had been Dave Leestma—he had been assigned as a mission specialist aboard 61-E, scheduled for March 1986, which was to focus various cameras and sensors on Halley's Comet. "Halley's Comet comes by every seventy-four years, and I get to go look at it," Dave said when he told me about it. "How about that?" A little later, Kathy Sullivan was assigned to 61-J, scheduled for August 1986, which was to deploy the space telescope. Later still, Jon McBride got a new assignment, not as pilot but as commander this time: he was to command 61-E, the Halley's Comet mission. Dave was to sit in the third seat, reading checklists for him, as he had done during 41-G. Doubtless Dave would have taken every opportunity to remind his new skipper not to dump the thruster fuel—in Jon's new status of veteran and commander, having graduated from 41-G with flying colors, Dave would have felt that Jon needed a few helpful reminders of his rookie days.

At NASA, old missions were quickly giving way to the new in a seemingly endless procession of flights. During my final visit, the next mission was in space—it was 51-A, whose primary task was to capture two satellites misfired into erratic orbits and to bring them back to earth. Even as I talked to the 41-G astronauts about their flight, we kept breaking off to listen to the 51-A air-to-ground conversations to see how the retrieval was coming.

I asked Dave if the trip had lived up to expectations. "Did it ever!" he said. "I haven't stopped smiling since I got back." I wanted to know if what they had said to Browder was true—that it was easier to talk about their trip to others who had been in space. "It's true that you can't describe it, and that it's best to talk with people who have been there," he said. "We just look at each other and nod." When I asked Kathy the same question a little later, she also agreed. "Yes," she said. "Much more can go unsaid. You don't have to search for words that you know will be inadequate anyway. When you say to another veteran astronaut, 'Remember what this or that was like?' you get a knowing nod. And then you know that the full experience is being shared." Even the photographs they took, Kathy felt, did not evoke being in space—at least, not to anyone who hadn't been there. If astronauts can talk only to other astronauts fully and completely about being in space, that may be another reason why they are such a tight group.

When I saw Browder, I asked him if he felt any sort of letdown, now that the flight was over. "There *was* a sort of postpartum depression after Crip's party—a feeling that we had built something, and now it had toppled over, and we had to build it again with a new crew. Of course, 41-G was the first mission our team had done together, and my first as a Team Lead. There's no mission that's ever like the first. I was talking to Dave about this at the party, and he agreed with me. He said he had felt that way all along, because of Sally. During our training, she was always talking about STS-7, *her* first mission. And John Fabian, who had been a mission specialist with her then, kept coming down to the simulators from time to time during our training sessions, to see how Sally was doing—there's a friendship that will last always. And that's the way we all now feel about each other, too. But time marches on, and we all get new missions."

The team had not been assigned to a new crew yet, though Ted had asked for 62-A, scheduled for January 1986. He had requested it because he thought it would be the first mission launched from the new facility at Vandenberg Air Force Base in California to go into polar orbit; he believed there was a good chance it would be assigned to Crippen because it was the sort of trail-blazing mission he usually got, and Ted and the team wanted to fly with Crippen again. While they were waiting to hear, the team was given the old 41-F crew, whose mission had been combined with 51-D's; they trained that crew for that new mission, launched the following year. At length, both Ted's prediction and his wish came true: Crippen was assigned the 62-A mission, and the 62-A mission was assigned to Browder's team. The only drawback was that it was a military flight, and therefore, Ted said, he would be unable to talk to anybody about it. I thought that would be a great hardship for Ted, and a great loss to his friends who profit by his comments. I particularly recalled one he had made an hour after the launch of 41-G, which, he said, "should have been scored musically." He had added, "I hope I always feel this way, for as long as I do, I know I will never want to do anything else, for it is this feeling that keeps us all going." In October 1984, enthusiasm was one quality Ted, and most others at the Space Center, never thought they would lose.

Epilogue

*T*hat overriding enthusiasm took quite a beating a year and a quarter—and twelve missions—after the 41-G crew had landed, when the *Challenger,* with the 51-L crew aboard, exploded during ascent on January 28, 1986. The accident occurred about seventy-three seconds after the launch, just after the moment of maximum dynamic stress while the solid rocket boosters were still firing and while the main engines were being brought back out of the bucket to full throttle. The last words received at Houston from the *Challenger* were "Go at throttle up." But a pair of seals in one of the joints in the right-hand booster failed, a jet of flame gushed from its side into the external tank, and 51-L ended in a conflagration.

My own view of the tragedy was very much influenced by what I knew of space flight from following the 41-G crew—and certain things about the 41-G mission became clearer in light of the accident, too. I remembered how, at the 41-G launch, Ted Browder and the rest of his team had not left the bleachers until main-engine cut-off eight and a half minutes into the mission, well after the booster separation. Now I understood the depth of their relief. I remembered, too, how John Young, then chief of the Astronaut Office, had said, "[The boosters] have to be a hundred percent reliable—you can't have a nozzle burn through, or have a seal go on you. The same is true of the main engines. . . . Liquid oxygen and liquid hydrogen are in close proximity to each other. Something could go." Something did—a booster seal, such as Young had mentioned. Young had also said, "The men and women who fly these things are very brave." That is a fact that the astronauts, the flight controllers, the instructors, and I myself had lost sight of in the course of the training for 41-G, because NASA and everyone in it, and many people observing it, already believed that space flight had become routine. In fact, that was my reason for writing this book: if space flights were becoming so routine that politicians, schoolteachers, and even journalists might go on them, we should see how one was put together.

In July 1986, six months after the accident, John Young told me, echoing the Report of the Presidential Commission on the Space Shuttle Challenger Accident, "Everyone thought we were operational, and we were not." We talked about a point he had made even before the 41-G mission, that most aircraft are tested hundreds of times before they are pronounced "operational" or are allowed to carry passengers who are not test pilots. Admittedly, there is no way a spacecraft can receive hundreds of such tests. Still, the space shuttle received only four test flights; by the fifth mission it was carrying mission specialists, and by the twelfth it was carrying nonastronauts. Astronauts have not often been happy about carrying nonastronauts aboard the space shuttle, for reasons in addition to the sort of in-groupiness exhibited by the 41-G crew. Young and others have suggested for some time that the shuttle had not flown enough to provide the confidence to carry nonessential passengers.

The 41-G mission, the thirteenth, was the second to carry payload specialists. When the *Challenger*'s boosters were retrieved from the Atlantic after the 41-G launch, no signs of scorching were detected on any of its O-rings (the seals in the joints)—though there had been scorching on the three preceding missions, 41-B, 41-C, and 41-D, and on all but two of the twelve that followed. One can only surmise that the 41-G mission had been in jeopardy, too, but had been very lucky.

Though the scorch marks on the previous missions were not brought up at the 41-G flight-readiness reviews, there were at the time many other symptoms of the underlying conditions that the Presidential Commission later identified as contributing to the tragedy. Many readers who are familiar with the report will already have picked them up for themselves. There are resonances almost everywhere: for one, the 41-G crew flew aboard the *Challenger* in order to allow the *Columbia* to have its ejection seats removed. Two years later, the commission would take NASA to task for not having incorporated some sort of crew-escape mechanism (though not necessarily ejection seats) into its orbiters. For another, at the time of 41-G, Crippen had already been waved off twice from landing at the Kennedy Space Center; even though he succeeded in landing there that time, the commission later recommended that there be no further landings at Kennedy until, among other things, there were better methods of predicting the weather there. And how often in the future, one wonders, will a satellite be deployed from the shuttle, now that the White House has ordered most of them to be launched on expendable rockets?

The pressure for the shuttle flights to be seen as routine, which the report stresses as a factor leading to the accident, was also beginning to build up in 1984, the year of 41-G, when there were six shuttle missions, more than in any previous year. The program, which NASA considered well along in the transition from a test program to an operational one, was gearing up to higher flight rates: nine in 1985 and fifteen planned for 1986. And 51-L was twelve missions and fifteen months after 41-G, which had represented a point midway on the graph leading to the explosion. The pressure for flights to be frequent and routine was the result of the need for the shuttle to be as economical as possible in order to carry commercial payloads. Consequently, NASA had to present the shuttle to the business community as being as reliable in its scheduling as the next carrier—in this case, *Ariane,* the European unmanned space launch vehicle. The 41-G mission was unusual in that it was made up almost entirely of NASA payloads, but it nonetheless suffered from scheduling problems, and simulator-time problems, that resulted from its being part of a long string of missions whose realities were dictated largely by marketing considerations. Indeed, in the wake of the 41-D postponement, 41-G with its NASA payloads was almost scrubbed in the interests of keeping commercial customers. When 41-E was scrubbed in its place, there was a rearrangement of payloads that caused a cascade of scheduling problems for the missions that followed. And the crew of 51-A, the flight that immediately followed 41-G, required a lot more time in the simulators while 41-G was still in preparation because it had been assigned its task of retrieving two lost satellites (in order to satisfy customers and insurance companies) extremely late. The Presidential Commission cited this situation as an example of NASA's "can-do" attitude, which, while commendable, in its view also contributed to the accident. When I wrote that there is a "universal tendency for schedules to slip," that propensity was just part of the problem. These scheduling pressures very likely contributed to the lateness of 41-G's mission-dedicated computer loads, and possibly even to the errors those loads contained. "Software tools and training facilities developed during a test program may not be suitable for the high volume of work in an operational environment," the commission found. "In many respects, the system was not prepared to meet an 'operational' schedule."

The way in which the pressure of training—greatly exacerbated by the scheduling pressures—tended to pile up toward the end of the 41-G mission was a situation that, according to the report, got worse on later missions. Ted Browder's concern that the 41-G mission might be the first ever to be delayed because the crew wasn't adequately trained turned out to be a common enough concern that the situation was stressed in the report. An astronaut, Henry Hartsfield, speaking of the period around the time of the accident, told the commission, "Had we not had the accident, we were going to be up against a wall. . . . For the first time, somebody was going to have to stand up and say we have got to slip the launch because we are not going to have the crew trained." This is almost a direct paraphrase of a comment of Ted Browder's in August 1984, two months before 41-G was launched. When I saw him again in July 1986, six months after the accident, he said, "I think every Team Lead goes through that feeling that he won't get his crew trained in time for the launch. Recently, I heard a manager say that if the tragedy hadn't occurred in the boosters, it might have happened in our area. Our biggest fear was that we could send up a crew not totally trained—though it would be trained to the best of our ability in the time allotted."

Ted concluded that it wasn't the will of the astronauts and their instructors, or their skill, that was lacking; it was the resources that they were given, as compared to the increasing pressure of the flight schedule. The Presidential Commission clearly agreed. Its eighth recommendation urged NASA to "establish a flight rate that is consistent with its resources." The White House recommendation in the summer of 1986 that NASA no longer accept commercial payloads, though a painful policy shift for many at NASA, would seem a step in the right direction, judging from the pressures that were already building up at the time of 41-G. And clearly the matter of whether a commander coming off another mission can join his crew late—under study during 41-G—will be less pressing with a lower flight rate.

Eugene Kranz, the director of Mission Operations, had told me before 41-G that in his opinion the flight controllers involved with the shuttle missions had not been "severely tested" in the manner of their Apollo counterparts, who on several occasions had had to make split-second life-and-death decisions. He told me six months after the accident that the *Challenger* explosion had done little to supply the sort of experience he had had in mind. It had not been a flight controllers' mission because there had been no warning of the problem, and hence nothing to work with. "The problem of that mission was that there were no options," he told me. "The flight controller's tradition is, 'Give us a few seconds, and we'll ace it.' But here there was no chance." It provided little experience useful to the flight controllers, or to the instructors.

Unlike the 41-D icicle, the 51-L explosion most likely will not find its way into sims. When, at the end of the 41-G contingency-training session that I had watched, Crippen said of the various aborts and ditches in the ocean that he had predicted, "I hope we don't have to do any of those hairy things," it was not a booster problem such as caused the 51-L accident that he had in mind. The instructors never bothered to simulate what would happen if a booster failed, because it was common knowledge that nothing could be done. Ted Browder told me after the accident that the instructors would never sim things that can't be solved. And Jack James, the propulsion instructor for integrated sims, who is now a Team Lead for stand-alones, told me, when I asked him whether the training shouldn't make it clearer that there *are* hopeless situations, "No. We want to teach the astronauts how to get out of tough situations. It doesn't do them any good to show them complete disasters. Our object is to push them to the limit of what they *can* get out of, and then to push that limit so that it grows and grows and grows. It doesn't teach them anything to show them something they cannot do." When in simulations the orbiter sometimes goes out of control, he added, the instructors stop the sim. I had seen Browder do that, sometimes at Crippen's request.

In view of the care that the instructors lavished on the astronauts during the 41-G training, I asked Ted Browder and some of his colleagues, six months after the accident, what they felt about the findings of the Presidential Commission—in particular its revelations that some engineers concerned with the development and certification of the boosters had not applied the same standards of excellence and logic to space flight that most people at the Johnson Space Center employed. The people involved in flying the shuttle—the astronauts and those who support them most immediately, the flight controllers and the instructors—were the survivors of the accident, as much as the families of those who died. At first, after the grief there was anger. "The anger was directed toward certain management people that didn't pass on essential information," Browder told me. From my experience observing the training for 41-G, the passing on of accurate information to those making the decisions (the astronauts, the flight controllers) was the cornerstone of space flight. The trust that those around you would do this was implicit in everything I had seen during the training of the 41-G crew. Steve Hamm, the computer instructor for the integrated sims, told me after the accident, "There was a sense of anger that such a thing could happen. There are a lot of conscientious people here, and all over NASA. It made us mad to read that those failures of the O-rings were discovered on previous flights and that they weren't reported and that the faulty equipment was not pulled off the line."

The Presidential Commission, in its fifth recommendation, came down very hard on NASA's need for better internal communication. The report was critical of engineers at the Marshall Space Flight Center in Huntsville, Alabama, who were in charge of the boosters, and it criticized that center for "a tendency . . . to management isolation" from the rest of NASA. The isolation, according to some NASA observers, had also led to an ingrown form of decision making. From my experience with the 41-G crew, the astronauts, flight controllers, and instructors at the Johnson Space Center, though something of an elite in-group, are the opposite of isolated: their whole professional life is a seething ferment of the exchange of ideas and the clash of opinions. Untransmitted ideas and convoluted thought processes would not survive for a minute in the environment of the simulators and the Mission Operations Control Center. Jack James said to me after the accident, "Here in operations, we go to meetings and we talk all the time—and if anyone here said something like 'Charred O-rings are all right,' ten guys would jump on him and say, 'That's hogwash.' " Perhaps, I thought, NASA should send Shannon O'Roark to Marshall with her light pen, in order to get the right people talking to each other.

Many of the people I had met in the course of 41-G were affected by the accident. Jon McBride and Dave Leestma and their 61-E flight had been next up after 51-L; they had been in their last couple of months of training for 61-E, which Jon would have commanded, when the *Challenger* was destroyed. Aside from their personal feelings at losing seven good friends, and also the spacecraft they had flown in before, it had been (as anyone who had followed 41-G might expect) a wrenching experience to stop training just as it was about to switch into the high gear of integrated sims. The training didn't stop instantly; it petered out over a month, so long as there was a possibility that flights might resume before too long. Their feelings were a mixture of great disappointment and relief. The cold weather on January 28, 1986, about thirty-six degrees at the time of launch and in the twenties the night before, had been an important factor in the accident; the coldest launch before that, 51-C, the previous January when the temperature had been fifty-three degrees, had suffered the most severe erosion of the O-rings before 51-L. The temperature was colder at the planned time of launch for 61-E. "We had been scheduled for a midnight launch on March 5," Dave said when I spoke to him six months after the accident. "That night, I was so disappointed that I could not sleep. About midnight, I called down to the Cape to see what the weather was. It was forty-seven degrees. So it could have been us."

The rest of the 41-G crew was affected, too. Sally Ride was a member of the Presidential Commission that investigated the accident, and Crippen—after serving as astronaut representative in the retrieval operations off the Cape—went on to chair a NASA committee to make recommendations for the structural and management changes within the space agency mandated by the commission. These had emphasized improved communications within the agency and the movement of astronauts and others with flight or operations experience into managerial positions. Following the commission's advice, NASA later appointed Crippen deputy director for space shuttle operations, which keeps him largely in Washington and at the Cape, and made John Young special assistant to the director of the Johnson Space Center for engineering, operations, and safety. (Both remain members of the astronaut corps and are eligible for further flights, though Young is no longer chief of the Astronaut Office.) Even before the commission's report was finished, Sally Ride became a special assistant to the administrator of NASA for long-range planning; the recommendations of her committee, entitled "Pathfinder," were to be issued in the summer of 1987 and reportedly would deal with space stations, moon bases, and bases on Mars. (In May 1987, she announced that she would leave NASA the following autumn to become a Science Fellow at Stanford University's Center for International Security and Arms Control.) At the time of the accident, Kathy Sullivan also was involved with long-range planning as a member of the National Commission on Space, which was examining what NASA should be doing over the next fifty years; the National Commission's report, published three months after the accident, is dedicated to the 51-L astronauts. Dave Leestma, no longer in training for 61-E after the accident, put his critical eye for hardware problems to good use as a member of a NASA committee investigating the evolution of the booster problems at Marshall; some of his ideas, he says, found their way into the report of the Presidential

Commission. In the summer of 1987, he, Kathy, and Jon McBride were still awaiting assignments to new crews. Their flight on 41-G seemed a very long time ago.

The instructors and flight controllers, and others close to operations, were hit almost as hard by the accident as the astronauts themselves. For a long time there was a feeling of what Ted Browder called "shared guilt," which he described as a concern on the part of himself and a good many of his friends that there might have been something that he or she or they could have done to avert the accident. Of course, there wasn't. Ted himself had had nothing to do with the training of the 51-L crew. Before the mission, he had been promoted to supervisor of all the payload-training instructors, while keeping his hand in at the consoles as a simulator instructor for the manipulator arm; but there was no arm aboard that flight, nor were payloads an issue. Besides, the Presidential Commission had found that "there was nothing that either the crew or the ground controllers could have done to avert the catastrophe." If the crew and the ground controllers had no responsibility for what happened, of course their instructors didn't either. Still, many people at the Johnson Space Center who had worked with the astronauts suffered from lingering doubt and depression, enough so that NASA and some of its contractors provided counseling for those who needed it. This is not surprising for a group who, as I knew full well, was so close and identified so wholeheartedly with its work.

Although Ted was not one of those who sought counseling, he told me he had been very dispirited for many months. He had known all the members of the 51-L crew and had worked with some of them, including Judy Resnik and Ellison Onizuka, on other flights. One day in May, four months after the accident, when he was still feeling the loss deeply, he saw an astronaut he knew showing two older women around the simulators. Ted said hello, and the astronaut introduced the women, who turned out to be the mother of Resnik and an aunt of Onizuka. "It was like someone had opened a door," Ted told me later. "I let it all pour out—I told them how much their children had meant to me, to all of us, and how we were going to miss them. I told them I was saying this as much for myself as for them—that their children represented a piece of my life I'll never change. They said they had heard all these things before from others, but each time it meant something new to them; they thanked me. When I left, I felt that the whole world had been lifted from my shoulders. No one could understand the depths of my feelings as they could. My sadness went away. I felt I could pick up the pieces and move on to the things they—the 51-L crew—would like."

In July 1986, Ted and the other instructors were busy with generic training and proficiency training for the pool of astronauts. In the manner of NASA, which is forever moving people into new combinations, Ted's old team had broken up. Shannon O'Roark and Randy Barckholtz have moved over to the integrated sims, where they are conspiring on nits, glitches, and malfs on an ever deeper level. Only Andy Foster remains on the team: he is now Team Lead. The simulators were running at one hundred ten hours a week instead of the hundred sixty hours they had been operating when the shuttles were flying. The pace, Ted thought, was just about right—enough to keep everyone very busy, but not enough to keep them in a permanent state of all-out effort. Ted and the others were looking forward to starting training for the first mission when flights resume. The crew for the next mission was appointed in the winter of 1987, though by the summer of 1987 their launch was at least a year away and no other crews had been named. Ted told me he suspects that when training starts in earnest once more, never again will it be as upbeat and light-hearted as it was at the time of 41-G. They will no longer have the same infectious confidence. An innocence has been lost.

Index

Abbey, George W. S., 21–26, 28, 245
abort(s), 115, 116–17
  contingency type, 115, 117, 120, 149
  RTLS type, 116–117, 118, 121, 122,
    202
  transatlantic, 117, 149–50
airplanes, 130–33, 135
  Gulfstream II, 130, 133, 135, 201
  KC-135, 29–30, 107, 135, 160
  T-38, 29, 130, 132–33, 204
air pollution, measurement, 17, 138,
    143
Allen, Joe, 152
anomalies. *See* malfunctions
Apollo (spacecraft)
  Apollo 13, 60, 119–20
  emergencies on, 119–20
  fire (1967), 5, 10
Apollo astronauts, and shuttle
    astronauts, compared, 19, 119–20
Arbet, Jim, 70, 72, 73, 74
arm. *See* manipulator arm
ascent(s) of orbiter, 115–16. *See also*
    abort(s)
  practiced, 135–36, 144–46, 201, 202,
    208–9
astronauts, 18–21
  assist other missions, 76, 155–56,
    235
  background and qualifications, 19
  bonding among, 24–25, 246
  characteristics of, 19–20; ambition,
    25–26; competitiveness, 25–26,

53–54, 98, 108, 123; economy of
    words, actions, 3–4; power of
    concentration, 84
crew selection, 18, 21–26
of mission 41-G. *See* crew of mission
    41-G
and mission planning, 16, 69, 77, 129
new, and Apollo astronauts,
    compared, 19, 119–20
oppose nonastronauts as passengers
    on flights, 153, 250
tasks and assignments, 20–21, 26
women. *See* women astronauts
attitude changes (of orbiter). *See also*
    orbital maneuvering system
  on flight of 41-G, 218
  scientific experiments and, 143

Barckholtz, Randell J. (Randy), 9, 49,
    79, 257
  background and career, 57
  in simulations, 57–58, 83, 91,
    97–98, 99, 147, 192
Bassham, Robert, 171, 174, 176–77,
    181, 213
Benson, Harold E., 127
bite(s), 83. *See also* malfunctions
blackout period(s), 98, 175, 194, 206,
    242
Brizzolara, William, 179
Browder, Ted H., 4, 8–9, 123. *See also*
    Team Leader
  background and career, 55–56

Browder *(continued)*
    and *Challenger* explosion, 253,
        256–57
    on characteristics of astronauts, 3–4
    on McBride, 42–44, 79, 144, 146,
        157, 225
    on planning scripts, 54–55
    and Rainey, 165–66, 173
    on readiness of crew for flight, 10,
        37–38, 83, 84, 93, 95, 125, 147,
        149, 157, 197
    restraining role on team, 52, 54, 60,
        85–86
    role in flight of 41-G, 207, 223, 242
bus(es). *See* data bus(es) for computers

cabin pressure, leaks in, 90, 91–93,
        189, 192
cameras, 29, 72–73, 150–51. *See also*
        photography
    IMAX, 27, 153–54
    large-format, 17, 138, 143
    movie, 151
    SPARKS, 18, 69, 142
    television, 151, 233
capsule communicator (CAPCOM), 14,
        20
cargoes (commercial) for shuttle
        flights. *See also* customers . . .
    affect flight plan and schedule,
        17–18, 65–68, 69, 142–43,
        178–79, 251
    NASA's assignment of, 16–17
    White House recommends
        eliminating, 252
Challenger (space shuttle)
    as craft for 41-G, 48, 81–82, 111
    explosion on mission 51-L, 249;
        assessment of, 249–56;
        consequences of, 5–6
    on 41-C mission, 77, 82
    repair and renovation, 82, 167
    return from Edwards AFB, 98
Chang, Franklin R., 208

changes in crew training, 2, 5, 60,
        101–2, 250–57
checklists and procedures manuals
    astronauts' relation to, 92, 95–96,
        121
    in simulations, 50, 66, 91–92, 167
    studying, 31, 43, 96
chores aboard orbiter, 70, 143, 159, 210
Collins, Michael: *Carrying the Fire*, 2
Columbia (space shuttle)
    for 41-G mission, 27, 48, 81–82
    renovated (1984), 17, 27, 81, 111
commander of 41-G mission. *See also*
        Crippen, Capt. R. L.
    and flight director, 141–42
    lack of, early in training. *See*
        Crippen, Capt. R. L.: absence . . .
    mission specialists vs. pilots as, 224
    and pilot, 43, 113, 144, 157, 225
    power and prerogatives of, 126–27
    and Sim Sup, 165–66, 173
commercial aspects of shuttle flight.
        See cargoes . . . ; customers . . .
communications loops, 35–36, 164–65,
        173, 176, 189
communications malfs, as "Catch-22,"
        187, 196, 216, 220
competition
    among astronauts, 25–26, 98
    among crews, 108
    within team, 98
    between team and crew, 53–54, 98;
        Crippen and, 123
computers, 34, 37
    Challenger vs. *Columbia* loads for,
        48, 138, 143, 202
    Crippen and, 22–23, 43, 123,
        147–48, 185, 193
    data buses for, 50–51, 58, 172
    malfs in, 40–41, 53, 83, 93–94,
        146–47, 187, 190–91, 195, 236,
        240
    for mission simulation, 34–35; malfs
        in, 74, 119, 149, 150
    redundancy in, 50–51, 94

stringing and restringing, 93–94, 147, 150

contingencies, training for, 10, 120–25, 213–15, 245

control rooms
    for flight controllers, 163–64
    for mission simulators, 35–36

cooling system
    flash evaporators, 220–23
    malfs in, 99, 147, 191
    radiator panels, 99, 147, 193

Cox, John T.
    and flight of 41-G, 214, 221, 234, 235
    and training for 41-G, 141–42, 166, 176–80, 185, 188, 196–97, 201

crashes and ditching, in simulations, 53, 100, 115, 117, 120

Creasey, Susan, 186

crew activity plan, 27, 68–70, 82, 141, 142–43, 166, 201
    and flight of 41-G, 209, 214, 218

crew of mission 41-G. *See also* decision making . . . ; flexibility . . . ; initiative . . . ; teamwork . . . ; trust . . . ; *and individual crew members*
    assignments: on 41-G mission, 26–27; after 41-G, 245
    and *Challenger* disaster, 254–55
    clothing and appearance, 26, 158
    "crystallizing" of, 80, 81, 111, 125, 147, 149, 157–58, 210
    curriculum, individualized, 29–30, 86
    and flight controllers, relationship with, 174–75, 176, 197, 225
    high spirits and joking, 37, 49, 75, 81, 88, 89, 101, 120, 182–83, 227–28
    liaisons with other operations, 27
    offices (Building 4), 28
    parties, 197–98, 244
    and payload specialists, relationship with, 125–26, 159, 160, 161, 226, 227, 244

prelaunch activities, 7–10, 201–4, 207–8
    quarantined, 201, 202, 204
    selection of, 18, 21–26
    and team of instructors, relationship with. *See* stand-alone team: and crew

crews of shuttle missions
    attention to ongoing flights, 76–77, 155
    bonding in, 24–25
    selection of, 18, 21–22
    variable membership, 33

crew stations, 34–35

Crippen, Capt. Robert L.
    absence of, from early training: advantages of, 41, 42, 79–80, 83–84, 145, 224; problems from, 49; Ride's role during, 24, 40, 48
    after *Challenger* disaster, 255
    assigned to 62-A mission, 247
    authority of, 42, 112, 177
    background and skills, 22–23
    competitiveness of, 123
    computer expertise, 22–23, 43, 123, 147–48, 185, 193
    and Cox, 141–42
    and crew of 41-G: communication, 48, 77, 111; his familiarity with, 25, 80; his leadership style, 3, 27, 121; initiative, 113; its selection, 23, 25; its teamwork, 33, 80, 111, 125
    on flight (mission 41-G), 223–24, 236, 240
    and flight controllers, 175
    joins training of 41-G in progress, 111–14, 125, 147–48, 150; integrated sims, 186–87, 193–94
    others' observations and opinions of: Browder, 95, 114, 223–24; Cox, 185–86; Hamm, 193–94; McBride, 111; O'Roark, 122–23; Ride, 27, 112; Sullivan, 80, 111, 112
    and Rainey, 165–66, 173

Crippen *(continued)*
  and team of instructors, 122–23,
      245, 246–47
  as veteran astronaut, 3, 7, 22, 80, 95,
      112; solves problems rapidly, 114,
      122, 185–86
customers for shuttle flights. *See also*
      cargoes . . .
  wooed by NASA, 18, 127–28, 140,
      251

dangers of space flight, 6, 12, 58–59,
      101, 200, 249
data buses for computers, malfs in,
      50–51, 58, 172
data dump, 119
debriefings
  of missions, 77
  of simulations, 49–50, 53, 59–60;
      of 3-day integrated, 196–97
decision making by crew, 37–38, 83,
      135, 174, 194
de-orbits. *See* descent(s) of orbiter
descent(s) of orbiter. *See also*
      landing(s) of orbiter
  emergency, 90–101, 189–97
  on flight of 41-G, 241
  practicing, 47–60, 90–101;
      integrated sims, 189–97, 202
Discovery (space shuttle), 82
ditching at sea, 115, 117, 120
Dream Is Alive, The (film), 153

earth-radiation budget satellite, 18
  deployment. *See* manipulator arm
  on flight of 41-G, 209, 210, 213, 214
  flight-readiness review of, 65–68
  in integrated sims, 167–68, 175,
      184–85
  photographs of, 142, 152
earth-watching functions of 41-G,
      17–18, 24, 179, 200, 219, 226. *See
      also* OSTA experiments
Edwards AFB, Calif., 82, 98, 133
Elachi, Charles, 218–19

electrical systems, 43, 57, 146, 196. *See
      also* MDM
Ellington AFB, Tex., 77, 130, 204, 243
emblem
  of crew, 77–78, 125
  of stand-alone team, 137
emergency de-orbits, 90–101, 189–97
Enterprise (space shuttle), 20
entry. *See* descent(s) of orbiter
EVA (extravehicular activity)
  on flight of 41-G, 231–36
  Leestma and, 27, 101–10, 230–36
  McBride and, 27, 30, 65, 182
  photo documentation of, 151, 152,
      233, 234
  practicing in WET-F, 101–10, 181,
      183, 198
  scheduling, 68, 69, 142, 230
  Sullivan and, 24, 27, 101–10; as first
      woman, 7, 181–82; on flight of
      41-G, 230–36
  tethering during, 107–8
experiments. *See* scientific
      experiments on 41-G

Fabian, John, 246
feature identification and location
      experiment, 17, 138, 143
Ferguson, Graeme, 153–54, 159, 230
fixed-base simulator, 35–42
flash evaporators, 220–23
flashlight, 105, 108–9, 198
flexibility
  of crew, 41, 73, 80, 224; as object of
      training, 213–15, 245
  of NASA, 214–15
  of scripts, 54, 55, 88, 214–15
flight controllers, 141. *See also* Mission
      Control
  and crew, relationship with, 174–75,
      176, 197, 225
  Crippen and, 175
  and flight of 41-G, 206–7
  at Goddard. *See* satellite controllers

and instructors, relationship with, 189
and integrated sims, 78, 112, 160, 171, 173–81; of EVA, in WET-F, 198; their location and duties, 163–64
officers and consoles, 164
shifts, 163, 184, 188; independence of, 222
turnover and training of, 163
flight data file, 201, 204. *See also* checklists and procedures manuals; crew activity plan
flight destination system, 23
flight director, 141, 177, 191. *See also* Rainey, E. Q.
flight of mission 41-G (5–13 Oct. 1984), 208–42
descent and landing, 241–42
EVA, 230–36
malfs, 210–25, 234–36
photography, 227–30, 241
and simulations, compared. *See* simulations . . . ; and space flight, compared
Flight Operations Review, 141–43
flight-readiness review(s), 65
of earth-radiation budget satellite, 65–68
of orbital refueling system, 126–30
food and meals in space, 69, 75, 89, 90, 159, 174, 210, 225
foot restraints (for EVA), 108, 109–10, 232, 233, 234
forced landing. *See* abort(s): contingency type
Foster, William A. (Andy), 49, 85, 95, 97, 99–100, 145–46, 242, 257
Fox, Alan, 144–49
Freehling, James R. (Jim), 187, 188, 190
Fullmer, Myron, 29, 30, 77, 78–79

Garneau, Cmdr. Marc, 7, 125, 143, 158, 159, 208, 210, 225–27, 243–44

Garza, Sonny, 170–77, 180–81
glitches. *See* malfunctions
Goddard Space Flight Center, Greenbelt, Md., its management of earth-radiation budget satellite
conservatism of, 178
on flight of 41-G, 209, 210, 214
and integrated sims, 178–79, 185
and NASA, decisions and negotiations, 65–68, 142
Griffin, Gerald D., 204
ground. *See* flight controllers; Mission Control; satellite controllers
Grubbs, Tom, 105, 108
Gulfstream II (aircraft), 130, 133, 135, 201

habitability classes, 29, 65, 160
Hail Columbia (film), 9, 153
Hamm, Stephen D., 169, 190, 193–94, 197, 254
Hartsfield, Henry W., Jr., 156, 252
Havens, Kitty, 183
Hilmers, David, 186–87, 209, 210, 242–43
Holmberg, William R.
and flight of 41-G, 206, 214, 215, 219
and training for 41-G, 68, 69–70, 82, 141, 143
Hughes, Francis, 16, 52
hydrazine fuel, 44, 110, 195
Crippen's concerns about, 69, 126–27, 128–29

icicles, 155–56, 182–83
IMAX camera, 27, 153–54, 159, 230
inertial-measurement units
alignment, 36–38, 40, 41–42, 185, 209
malfs in, 51–52, 145, 169
initiative of crew, 113, 135, 174, 235
instructors. *See* integrated instructors; Johnson Space Center: organization . . . ; stand-alone team

integrated instructors, 173–81
  relation to: crew, 136, 189; Crippen, 185–86; flight controllers, 189; stand-alone team, 136, 173
  script-writing session, 167–72
integrated simulations, 78–79, 155–99
  ascents, 135–36
  astronaut errors in, 181, 185
  of Day 1, 173–81
  of Day 5 and 6, 201
  first after Crippen's return, 78–79, 111–112
  and flight, compared, 206, 207, 208, 210, 221
  purposes of, 78, 160, 166, 169
  scripts and scenarios for, 135–36, 166; writing, 165–72, 178
  and stand-alones, compared, 135, 160, 166–67, 174, 175, 184, 194
  three-day, 144, 184–96, 218; debriefing, 196–97; script for, 167–72
  verisimilitude of, 179, 181

James, John E. (Jack), 169–70, 178, 180, 188, 189, 190, 194–95, 198, 253, 254
Johnson Space Center, Houston, 15–16
  buildings, 28, 34
  organization and divisions of, 16, 27–28

KC-135 (aircraft), 29–30, 107, 135, 160
Kennedy Space Center, Cape Canaveral, Fla. (landing orbiters at), 81–82, 111, 116–17, 241, 250
Keppler, Carl, III, 170–71, 189–90, 191–92, 195, 222
Kranz, Eugene, 48, 55, 60, 119–20, 252–53
Ku-band antenna, 148, 215, 216, 218, 219, 234–36

landing gear, malfs in, 48–49
landing(s) of orbiter. *See also* descent(s) of orbiter

crosswinds and, 100, 242
  forced. *See* abort(s): contingency type
  on mission 41-G, 241–42
  simulation and practice, 100; in aircraft, 29–30, 132–34, 204
  sites for, 82, 133; for emergency, 92, 189, 192; for 41-G, 111
  steepness and velocity, 132–33, 241
Landsat (satellite), 102, 128
launch date for 41-G, 82, 111, 139–40, 150, 252
launch of 41-G (5 Oct. 1984), 12–14, 208
LDEF (long-duration exposure facility), 73–75, 76
leaks, 45–46, 53, 85, 87, 145
  in cabin pressure, 90, 91–93, 189, 192
Leestma, Cmdr. David C., 24, 69–70
  assignments after 41-G, 245, 255–56
  background and character, 72
  and *Challenger* disaster, 254
  and design of tools, 108–9, 129–30
  and EVA, 27, 70–75, 101–10, 230–36
  on flight of 41-G, 224, 246
  as rookie astronaut, 7, 77
  in training sessions, 70–75, 101–10, 181, 183, 198
LOC (loss of control), 151
loss-of-signal periods, 98, 175, 194, 206, 242
Lousma, Jack, 152

McBride, Jim, 106, 108, 234
McBride, Cmdr. Jon A. *See also* pilot of 41-G mission
  assignment, after 41-G, 245, 256
  background, career, and personality, 43–44
  Browder on, 42–44, 79, 144, 146, 157, 225
  and *Challenger* disaster, 254

and Crippen, 43, 113, 144, 157, 225
and EVA, 27, 30, 65, 182
on flight of 41-G, 225
as liaison with planners, 27, 69
and photography, 151, 153, 230
progress of, in training process, 79,
    86, 113, 144, 146, 157, 194, 225
as rookie, 7, 23, 42, 43, 79
in training sessions, 36–60, 65, 83,
    85, 87, 88, 96–97, 130–34,
    144–46, 185, 186, 201, 208–9
McLeaish, John E., 2, 3
Mailer, Norman: *Of a Fire on the
    Moon,* 2
main-engine cut-off (MECO), 14, 116,
    208
maintenance of equipment, inflight,
    27, 216
maintenance tasks, 70, 143, 159, 210
malfunctions (malfs, glitches, nits,
    anomalies, bites). *See also under
    specific equipment*
  complex combinations, 51, 156, 169,
    175, 184, 194, 200
  as "farfetched," 60, 236, 240
  on flight of 41-G, 210–25, 234–36
  in integrated sims, 165, 168–72, 184
  repeated, for practice, 59, 86, 175
  role in training, 32, 213–14
manipulator arm, 70–71
  deploys earth-radiation budget
    satellite, 67–68, 214; practicing,
    70–75, 138, 177–78, 180, 184–85
  malfs in, 172, 175–76, 177, 191
  Ride and, 38, 67–68, 70–75
  single-joint mode, 71, 172
  training with, 29, 70–75, 138
manipulator-development facility, 29,
    70–75, 138
  and mission simulator, 74, 75
manned maneuvering unit, 76
Maximum Q, 120–21
MDM (multiplexer-demultiplexer), 57,
    144
meals. *See* food and meals in space

medical experiments, 159, 226
Mission Control. *See also* flight
    controllers
  communications with, 93; loss-of-
    signal periods, 98, 175, 194, 206
  and crew, relationship with, 174–75
  at Houston, for 41-G, 111
Mission Operations Control Center, 141
  computers of, 35
  Simulation Control Area, 164–65,
    188
Mission Operations Control Room,
    163–64, 165
mission rules, 141
  flight director and, 141, 177
  and flight of 41-G, 216, 217, 231
  tested in integrated sims, 163, 166,
    175, 176, 178, 180–81, 191, 194
missions. *See* shuttle missions
mission simulators. *See* shuttle
    mission simulators
mission specialist astronauts, 18–19,
    20, 144–45, 224
motion-based simulator, 35, 117–25

NASA (National Aeronautics and
    Space Administration)
  and cargoes and customers for space
    flights, 16–17, 18, 65–66, 127–28,
    140, 251
  criticism of, and changes in, after
    Challenger disaster, 249–56
  Office of Space Science and
    Applications (OSSA), 17
  protects crews from publicity, 2–3
  training philosophy, xii, 33, 214–15
National Commission on Space, 255
navigation-state vector, 51
Nelson, George, 76, 108, 110, 235
Newman, Frank, 204
nits, 41. *See also* malfunctions
nominal training, 156

obscenities, 91
ocean, ditching in, 115, 117, 120

one-G trainer, 29, 70, 151. *See also* manipulator-development facility
optimism and enthusiasm at NASA, 6, 200, 247, 249, 257
orbit(s) of 41-G, 146, 170
  scientific experiments and, 17–18, 66–67, 142–43
orbital activities, 146–49, 209
orbital maneuvering system, 44, 68, 85, 93–94, 116, 146, 170, 194, 195
orbital refueling system, 18, 140
  in EVA training, 102, 105–10
  on flight of 41-G, 209, 232–34
  Leestma and, 24, 27
  photographs of, 151, 152, 233, 234
  readiness review of, 126–30
orbital skills training, 82–89
orbiter(s), 12–13. See also *Challenger; Columbia; Discovery; Enterprise*
  selection, for 41-G, 27, 48, 81–82, 111
orientation of spacecraft, 36–38, 40, 41–42
O'Roark, Shannon L., 36, 50–51, 53, 58, 83, 93, 99, 122–23, 136, 257
  background, character, and appearance, 51–52
OSTA experiments, 17, 27, 66, 67, 138, 143, 169

pace of training, 112, 135 174, 256–57
  quickens, 77, 138, 157, 200, 251–52
  Team Leader controls, 83, 93
payload controllers. *See* satellite controllers
payload requirements
  and orbit and attitude decisions, 65–68, 178–79
  and schedule, 69, 142–43, 251
payloads control center. *See* Goddard Space Flight Center, Greenbelt, Md.
payload specialists, 7, 81, 125–26, 174, 197. *See also* Garneau, Cmdr. M.; Scully-Power, P.

and crew, relationship with, 125–26, 159, 160, 161, 226, 227, 244
  on flight of 41-G, 210, 225–27
  help with chores, 159, 210, 225
  and press, 227, 243–44
  training of, 160, 210, 227
photography. *See also* cameras
  "antic," 152–53, 227–28
  on flight of 41-G, 227–30, 241
  training in, 29, 150–54
pilot astronaut(s), 18–19, 20, 130–33
  as commander of mission, 224
pilot of 41-G mission, 27, 79. *See also* McBride, Cmdr. J.
  and commander, 43, 113, 144, 157, 225
  selection of, 23–24
  training in aircraft, 29–30, 132–34, 204
planning of missions. *See also* Flight Operations Review; flight-readiness review(s); schedule(s)
  actual missions' effect on, 223
  astronauts' influence on, 16, 69, 77, 129
  payload requirements and, 17, 65–68, 69, 142–43, 178–79, 251
powerdown of instruments, 58, 99, 191, 196
prebreathing, 182, 231
Presidential Commission on the Space Shuttle Challenger Accident: *Report,* 5, 250–56
press conferences, 158–59, 202, 227, 243–44
pressure
  on crew, 157, 251–52
  on NASA, by radar scientists, 179–80
  by NASA, for routine, frequent shuttle flights, 251–52
procedures manuals. *See* checklists and procedures manuals
publicity
  astronauts and, 2–3, 158–59, 202

payload specialists and, 227, 243–44
pumps, malfs in, 37, 87, 97–98,
  187–88, 189, 191

quarantine of crew, 201, 202, 204

radar, shuttle-imaging (SIR-B), 17,
  142–43
  on flight of 41-G, 216–18, 231
  integrated sims and, 168, 179–80,
    186–87
  Sullivan and, 24, 27, 138, 216–18
radiator panels, 99, 147, 193
Rainey, Edwin Q., 165–69, 173–81,
  187, 191, 197, 221
reaction-control system, 44–45
readiness reviews. *See* flight-readiness
  review(s)
Reagan, Ronald, 240
redundancy
  in computers, 50–51
  personal, of crew, 41
  in thruster system, 87, 209
refueling system for satellites. *See*
  orbital refueling system
Regency trainer, 31
regimbaling, 94
Remote Payload Operations Control
  Center, 66. *See also* Goddard
  Space Flight Center, Greenbelt,
  Md.
renovation of orbiters
  of *Challenger,* 82, 167
  of *Columbia,* 17, 27, 81, 111
  effect on flight planning, 17
repairs. *See also* malfunctions
  of equipment, maintenance in flight,
    27, 216
  of satellites, 23, 76, 110
Richmond, Mel, 136, 160
Ride, Sally K.
  assists other missions during flight,
    76, 155–56
  assists others in crew, 40, 65, 72
  and *Challenger* disaster, 255

education, career, appearance, and
  personality, 40
on flight of 41-G, 216
and manipulator arm, 38, 67–68,
  70–75
prior experience as astronaut, 7, 24;
  on STS-7, 23, 38, 72
role of, in Crippen's absence from
  training, 24, 40, 48
in training sessions, 38, 70–75
risks. *See* dangers of space flight
rivalry. *See* competition
Ross, Maj. Jerry L., 235
RTLS (return to launch site)
  landing(s), 116–17, 118, 121, 122,
    202
rules. *See* mission rules
Russell, Bill, 48–49, 50, 53

satellite controllers (at Goddard), 65
  and flight of 41-G, 209, 210, 214
  and integrated sims, 178–79, 185
satellites. *See also* earth-radiation
  budget satellite; Landsat; Solar
  Max; SPAS
  refueling. *See* orbital refueling
    system
  repairing and servicing, 23, 76, 110,
    127–28
  tracking and data-relay type, 66,
    219–20
Savitskaya, Svetlana, 181–82
schedule(s). *See also* crew activity plan
  daily/weekly, for astronauts, 30–31
  for EVA, 68, 69, 142, 230
  for launch date, 82, 111, 139–40, 150,
    252
  NASA's pressure on, 251–52
  payload requirements and, 69,
    142–43, 251
  for three-day sim, 167
  for training, getting behind in, 150,
    156, 157–58
Schmitt, Harrison, 24

scientific experiments on 41-G, 17–18, 138
  Canadian, 159
  on flight: payload specialists and, 226–27; revised after malfs, 218–19
  medical, 159, 226
scripts for simulations. *See* integrated simulations: scripts and scenarios for; stand-alone team: scripts and scenarios of
Scully-Power, Paul, 7, 125, 158, 159, 174, 210, 225–27, 243
sensor(s), malfs in, 83, 85, 157, 173–74, 192–93
Seus, Stephen, 151–52, 230, 241
shuttle crew. *See* crew of mission 41-G; crews of shuttle missions
shuttle-imaging radar. *See* radar, shuttle-imaging (SIR-B)
shuttle mission 41-G, 5
  crew. *See* crew of mission 41-G
  flight of. *See* flight of mission 41-G
  launch (5 Oct. 1984), 12–14, 208
  postflight activities, 243–45
shuttle missions
  commercial aspects. *See* cargoes . . . ; customers . . .
  frequency, 2, 5, 20, 251
  in orbit, attention to by other crews, 76–77, 155
  planning. *See* planning of missions; schedule(s)
shuttle missions, by number, 23
  41-C (formerly STS-13), 23, 73, 76–77, 98, 113, 152
  41-D, 12, 82, 139–40, 155–56
  41-E, 139, 140, 251
  41-F, 82
  41-G. *See* shuttle mission 41-G
  51-A, 245–46, 251
  51-L (explosion, Jan. 1986), 5, 249
  61-E, 245, 254
  61-J, 245
  62-A, 247

STS-7, 23, 38, 72, 181
STS-9, 25
STS-11, 110
shuttle mission simulators, 30, 31–32, 34–36, 74–75, 82–89, 144–49
  and aircraft, for training in landings, compared, 135
  cost of, 35
  crew stations, 34–35
  fixed-base, 35–42
  and manipulator-development facility, relative advantages, 74, 75
  motion-based, 35, 117–25
shuttle orbiter. *See* orbiter(s)
sim crash(es), 119, 149, 150
simulations of space-flight conditions, 32–33
  increasing complexity and length of, 41, 144, 156, 169, 175, 184, 194, 200
  integrated. *See* integrated simulations
  scripts and scenarios for. *See* integrated simulations: scripts and scenarios for; stand-alone team: scripts and scenarios of
  and space flight, compared, 213–14; ascents, 209; crew and, 207, 208; emotional level, 218–19; flight controllers and, 206; malfs (identical) occurring in both, 213–14, 217, 221, 222, 244; WET-F, verisimilitude of, 232
  verisimilitude of, 36, 179, 181
Simulation Supervisor (Sim Sup), 135–36, 164, 165
simulators. *See* one-G trainer; Regency trainer; shuttle mission simulators; single-system trainer; WET-F
single-system trainer, 31, 44
Solar Max (satellite), 23, 76, 102, 108, 110, 235
solotons (oceanic), 159, 226

spacesuit(s), 105–6, 200, 232
  classes in, 29, 65
Space Transportation System (STS), 17
  flight designations, 23
space walk. *See* EVA (extravehicular
    activity)
SPARKS (earth-observing camera), 18,
    69, 142
SPAS (satellite), 23, 38, 72, 181
speed brakes, 100, 133
spiral eddies (oceanic), 159, 226
stand-alone team, 4. *See also*
      Barckholtz, R.; Browder, T.;
      Foster, W. A.; O'Roark, S. L.
    Browder's restraint on, 52, 54, 60,
      85–86
    changes in personnel, 49
    competition among, 98
    and crew, relationship with:
      affection and loyalty, 86, 202,
      204–5, 243, 246; balance needed
      in, 52; competition, 32, 53–54, 98;
      continuity/stability, 49, 86–87; as
      coordinated unit, 89; and Crippen,
      122–23, 246–47; in integrated
      sims, 136–37
    Crippen on, 245
    high-spirited moods of, 9–10, 52
    at landing of 41-G, 242, 243
    at launch of 41-G, 9–10, 13–14,
      204–5
    role in flight of 41-G, 207
    role in integrated sims, 136–37, 164,
      166, 173
    scripts and scenarios of: flexibility
      in, 54, 55, 88; improvising, 40–41,
      88, 98, 145; and integrated sim
      scripts, compared, 166–67; vs.
      other teams' style, 145; planning,
      54–55, 56–57; reusable
      (proposed), 55
    work area of, 35–36
stand-alone training. *See also*
    malfunctions; stand-alone team
    definition, 32

  and integrated sims, compared, 135,
    160, 166–67, 174, 175, 184, 194
  purpose, 79
stars, alignment of spacecraft by,
    36–42, 89
Stewart, Col. Robert L., 105, 108, 110
Suarez, Mark, 171
Sullivan, Kathryn D. (Kathy), 24, 179
  appearance and personality, 84, 92
  assignments, after 41-G, 245,
    255–56
  background, 84, 138–39
  on checklists, 92, 96
  and EVA, 24, 27, 101–10,
    230–36
  on flight of 41-G, 216–18, 219,
    230–36, 246
  and OSTA experiments, 27, 138
  and photography, 151, 234, 241
  as rookie astronaut, 7, 84
  in training sessions, 84, 101–10,
    148–49, 181, 183, 198
survival training, 102

TACAN (tactical air navigation
    system), malfs in, 54, 58–60, 99
talk-backs, 173–74
Team Leader, 32. *See also* Browder, T.
  controls training pace, 83, 93
  restraining influence on team, 52,
    54, 60, 85–86
  and Sim Sup, 165–66, 173
teamwork of crew, 33, 41, 80, 125, 235
tiles, thermal, 82, 97, 98, 167
tip-off rates, 73
Todd, B. Keith, 70, 170, 172, 176, 180,
    187, 191, 192–93, 195, 242
toilet, 65, 125–26, 155, 221
tools used in space, 105, 107, 232, 233
  design of, 108–9, 129–30
trajectory of 41-G, 17–18. *See also*
    orbit(s) of 41-G
transatlantic aborts, 117, 149–50
trust among 41-G crew, 33, 113
T-38 (aircraft), 29, 130, 132–33, 204

umbilical for EVA, 171–72, 175, 177,
    184

Van Hoften, James D., 76, 106, 152
verniers, 44, 87, 180
versatility of crew. *See* flexibility: of
    crew

waste-management system, 65,
    125–26, 155, 221
weightlessness, 209. *See also* WET-F
    simulation of, 70, 106–7, 135
Weitenhagen, Ronald A. (Ron), 169,
    176, 217
WET-F (weightless environment
    training facility), 20, 29, 101, 181,
    183
    accidents in, 106
    EVA simulation in, 101–10, 183, 198;
        and flight situation, compared,
        232

Williams, Steve, 118–24
windows of orbiter
    access to, 125, 226, 227, 241
    views from, 35, 36, 68, 117–18, 241
Wolfe, Tom: *Right Stuff, The,* 2
women astronauts, 18
    on EVAs, 7, 181–82
    on 41-G crew, 24
    Savitskaya, 181–82
    status at NASA, 28–29
    susceptibility to bends, 182

Young, John
    after *Challenger* disaster, 255
    commands STS-9, 25
    and crew selection, 21–22
    on risks of space flight, 12, 249
    on teamwork, 80

Zweig, Roger, 130, 132, 135

HENRY S. F. COOPER, JR., is a staff writer for *The New Yorker,* where his coverage of the space program has appeared for two decades. Many of his articles have appeared in book form, including *Apollo on the Moon, Moon Rocks, Thirteen: The Flight That Failed, A House in Space, The Search for Life on Mars,* and *Imaging Saturn.*

Designed by Martha Farlow.

Composed by Brushwood Graphics Inc., in Century Schoolbook (condensed) with Futura Bold Italic.

Printed by R. R. Donnelley & Sons Company on 50-lb. Cream-White Antique and bound in Joanna Kennett and Rainbow-2 Dublin.